Nutritional Biochemistry

Nutritional Biochemistry

Dr. B.K.P. Singh

AMIGA PRESS. INC
Delhi -110051

AMIGA PRESS. INC
H.O :- 28-A, Indra Park, Chander Nagar
Delhi :- 110051 (India)
Tel :- 011-22504028, +91-98736-57000
Email :- amigapress@gmail.com

This edition is published by Amiga Press.Inc

ISBN : 978-93-84533-18-2

First Published 2018

Printed and bound in India by Thomson Press Ltd., New Delhi

Preface

A nutrient is a chemical that an organism needs to live and grow or a substance used in an organism's metabolism which must be taken in from its environment. Nutrients are the substances that enrich the body. They build and repair tissues, give heat and energy, and regulate body processes.

An essential nutrient is a nutrient required for normal body functioning that either cannot be synthesized by the body at all, or cannot be synthesized in amounts adequate for good health (e.g. niacin, choline), and thus must be obtained from a dietary source. Essential nutrients are also defined by the collective physiological evidence for their importance in the diet, as represented in e.g. US government approved tables for Dietary Reference Intake.

Some categories of essential nutrients include vitamins, dietary minerals, essential fatty acids, and essential amino acids. Water and oxygen are also essential for human health and life, as oxygen cannot be synthesized by the body, and water, while a biochemical reaction product of metabolism, is not created in sufficient amounts. Both are necessary as biochemical reactants in some processes, and water is used in various ways such as a solvent, carrier, coolant, and integral polar structural member, but both are often not included as nutrients.

Different species have very different essential nutrients. For example, most mammals synthesize their own ascorbic acid, and it is therefore not considered an essential nutrient for such species. It is, however, an essential nutrient for human beings, who require external sources of ascorbic acid (known as Vitamin C in the context of nutrition).

Carbohydrates are compounds made up of sugars. Carbohydrates are classified by their number of sugar units: monosaccharides (such as glucose and fructose), disaccharides (such as sucrose and lactose), oligosaccharides, and polysaccharides (such as starch, glycogen, and cellulose).

Protein requirement is also increased during lactation due to the milk production. The Indian council of medical research has recommended an additional daily intake of 25gm for the first six months and 18 gm during the six to twelve months of lactation. If the protein and energy is lacking in the mothers diet there will be reduction in milk volume rather than in milk quality but if very low protein are taken then the proportion of casein may be reduced. The extra protein can be achieved by including protein rich foods like milk and milk products, egg, meat, fish, poultry, nuts, and cereal and pulse combinations.

Calcium is one of the important minerals required both during pregnancy and during lactation. During pregnancy it is required for the foetal skeletal development and during lactation for the milk production. The first three months of the baby is fed only on the breast milk so again the baby depends on the mother for the nutrient requirement. An additional 1000mg per day calcium is required during this period.500 ml milk or milk products per day must be taken for the calcium requirement and the remaining from the other food stuff like ragi, Bengal gram whole, soyabean, amaranth, fenugreek leaves and fish.

The uniqueness of this book hence lies in the editor's way of reconstructing the chapter under review by delving deep into the areas of the subject.

—Editor

Contents

1

Introduction

NUTRIENT

A nutrient is a chemical that an organism needs to live and grow or a substance used in an organism's metabolism which must be taken in from its environment. Nutrients are the substances that enrich the body. They build and repair tissues, give heat and energy, and regulate body processes. Methods for nutrient intake vary, with animals and protists consuming foods that are digested by an internal digestive system, but most plants ingest nutrients directly from the soil through their roots or from the atmosphere. Some plants, like carnivorous plants, externally digest nutrients from animals, before ingesting them. The effects of nutrients are dose-dependent.

Organic nutrients include carbohydrates, fats, proteins (or their building blocks, amino acids), and vitamins. Inorganic chemical compounds such as dietary minerals, water, and oxygen may also be considered nutrients. A nutrient is essential to an organism if it cannot be synthesized by the organism in sufficient quantities and must be obtained from an external source. Nutrients needed in large quantities are called macronutrients; micronutrients are required in only small quantities.

TYPES OF NUTRIENT

Macronutrients are defined in several different ways.

The chemical elements humans consume in the largest quantities are carbon, hydrogen, nitrogen, oxygen, phosphorus, and sulfur. The classes of chemical compounds humans consume in the largest quantities and which provide bulk energy is carbohydrates, proteins, and fats. Water and atmospheric oxygen also must be consumed in large quantities, but are not always considered "food" or "nutrients".

Calcium, salt (sodium and chloride), magnesium, and potassium (along with phosphorus and sulfur) are sometimes added to the list of macronutrients because they are required in large quantities compared to other vitamins and minerals. They are sometimes referred to as the macrominerals.

The remaining vitamins, minerals, fats or elements, are called micronutrients because they are required in relatively small quantities.

Substances that Provide Energy

Carbohydrates are compounds made up of sugars. Carbohydrates are classified by their number of sugar units: monosaccharides (such as glucose and fructose), disaccharides (such as sucrose and lactose), oligosaccharides, and polysaccharides (such as starch, glycogen, and cellulose).

Proteins are organic compounds that consist of the amino acids joined by peptide bonds. The body cannot manufacture some of the amino acids (termed essential amino acids); the diet must supply these. In nutrition, proteins are broken down through digestion by proteases back into free amino acids.

Fats consist of a glycerin molecule with three fatty acids attached. Fatty acids are unbranched hydrocarbon chains, connected by single bonds alone (saturated fatty acids) or by both double and single bonds (unsaturated fatty acids). Fats are needed to keep cell membranes functioning properly, to insulate body organs against shock, to keep body temperature stable, and to maintain healthy skin and hair. The body does not manufacture certain fatty acids (termed essential fatty acids) and the diet must supply these.

Fat has an energy content of 9 kcal/g (~37.7 kJ/g); proteins and carbohydrates 4 kcal/g (~16.7 kJ/g). Ethanol (grain alcohol) has an energy content of 7 kcal/g (~29.3 kJ/g).

Substances that support metabolism:

- Dietary minerals are generally trace elements, salts, or ions such as copper and iron. Some of these minerals are essential to human metabolism.
- Vitamins are organic compounds essential to the body. They usually act as coenzymes or cofactors for various proteins in the body.
- Water is an essential nutrient and is the solvent in which all the chemical reactions of life take place.

The strip of a green alga (*Enteromorpha*) along this shore indicates that there is a nearby source of nutrients (probably nitrates or ammonia from a small estuary).

The chemical elements consumed in the greatest quantities by plants are carbon, hydrogen, and oxygen. These are present in the environment in the form of water and carbon dioxide; energy is provided by sunlight. Nitrogen, phosphorus, potassium, and sulfur are also needed in relatively large quantities.

Together, these are the elemental macronutrients for plants, often represented by the acronym CHNOPS. Usually they are sourced from inorganic (e.g. carbon dioxide, water, nitrate, phosphate, sulfate) or organic (e.g. carbohydrates, lipids, proteins) compounds, although elemental diatomic molecules of nitrogen and (especially) oxygen are often used. Other chemical elements are also necessary to carry out various life processes and build structures.

Some of these are considered macronutrients in certain organisms. The mnemonic C. HOPKiN'S CaFe Mg (to be used as C. Hopkins coffee mug) is used by some students to remember the list as: carbon, hydrogen, oxygen, phosphorus, potassium, nitrogen, sulfur, calcium, iron, and magnesium. Silicon, chloride, sodium, copper, zinc, and molybdenum are sometimes also included, but are in other cases considered micronutrients.

Oversupply of plant nutrients in the environment can cause excessive plant and algae growth. Eutrophication, as this process is called, may cause imblances in population numbers and other nutrients that can be harmful to certain species. For example, an algal bloom can deplete the oxygen available for fish to breathe. Causes include water pollution from sewage or runoff from farms (carrying excess agricultural fertilizer). Nitrogen and phosphorus are most commonly the limiting factor in growth, and thus the most likely to trigger eutrophication when introduced artificially.

Essential Nutrient

An essential nutrient is a nutrient required for normal body functioning that either cannot be synthesized by the body at all, or cannot be synthesized in amounts adequate for good health (e.g. niacin, choline), and thus must be obtained from a dietary source. Essential nutrients are also defined by the collective physiological evidence for their importance in the diet, as represented in e.g. US government approved tables for Dietary Reference Intake.

Some categories of essential nutrients include vitamins, dietary minerals, essential fatty acids, and essential amino acids. Water and oxygen are also essential for human health and life, as oxygen cannot be synthesized by the body, and water, while a biochemical reaction product of metabolism, is not created in sufficient amounts. Both are necessary as biochemical reactants in some processes, and water is used in various ways such as a solvent, carrier, coolant, and integral polar structural member, but both are often not included as nutrients.

Different species have very different essential nutrients. For example, most mammals synthesize their own ascorbic acid, and it is therefore not considered an essential nutrient for such species. It is, however, an essential nutrient for human beings, who require external sources of ascorbic acid (known as Vitamin C in the context of nutrition).

Many essential nutrients are toxic in large doses. Some can be taken in amounts larger than required in a typical diet, with no apparent ill effects.

Linus Pauling said of vitamin B_3, (either niacin or niacinamide), "What astonished me was the very low toxicity of a substance that has such very great physiological power. A little pinch, 5 mg, every day, is enough to keep a person from dying of pellagra, but it is so lacking in toxicity that ten thousand times as much can [sometimes] be taken without harm."

Fatty acids:

o á-Linolenic acid (the shortest chain omega-3 fatty acid)

o Linoleic acid (the shortest chain omega-6 fatty acid).

Amino acids:

- Isoleucine
- Lysine
- Leucine
- Methionine
- Phenylalanine
- Threonine
- Tryptophan
- Valine.

Essential amino acids necessary for human children but not adults:

- Histidine
- Arginine.

Vitamins:

- Vitamin A (retinol)
- Vitamin Bp (choline)
- Vitamin B1 (thiamin)
- Vitamin B2 (riboflavin, vitamin G)
- Vitamin B3 (niacin, vitamin P, vitamin PP)
- Vitamin B5 (pantothenic acid)
- Vitamin B6 (pyridoxine, pyridoxamine, or pyridoxal)
- Vitamin B7 (biotin, vitamin H)
- Vitamin B9 (folic acid, folate, vitamin M)
- Vitamin B12 (cobalamin)
- Vitamin C (ascorbic acid)
- Vitamin D (ergocalciferol, or cholecalciferol)
- Vitamin E (tocopherol)
- Vitamin K (naphthoquinoids).

Dietary minerals:

- Calcium (Ca)

- Chloride ($Cl^{''}$)
- Chromium (Cr)
- Cobalt (Co) (as part of Vitamin B-12)
- Copper (Cu)
- Iodine (I)
- Iron (Fe)
- Magnesium (Mg)
- Manganese (Mn)
- Molybdenum (Mo)
- Nickel (Ni)
- Phosphorus (P)
- Potassium (K)
- Selenium (Se)
- Sodium (Na)
- Sulfur (S) numerous roles
- Zinc (Zn).

The required quantity varies widely between nutrients. At one extreme, a 70 kg human contains 1.0 kg of calcium, but only 3 mg of cobalt.

DIETARY STUDIES

Object of Dietary Studies

The quantity of food which different families purchase varies between wide limits; a portion being lost mechanically in preparation and a still larger and more variable amount in the refuse and non-edible parts. If a record is made of all foods purchased and the waste and non-edible portions are deducted, the nutrients consumed by a family may be calculated by multiplying the weight of each food by the average composition. If such calculations be made, it will be found that in some families nearly a half pound per day of both protein and fat is consumed by adults, while in other families less than half of this amount is used. The object of dietary studies is to determine the source, cost, composition, and nutritive value of the foods consumed by different families; they also enable comparisons to be made of the amounts of nutrients purchased. Extensive dietary studies have been made by the United States Department of Agriculture, and the results have been published in various bulletins.

Wide and Narrow Rations: When the amount of carbohydrates in a ration is small in comparison with the protein, it is called a narrow ration, while a wide ration is one in which the carbohydrates are much in excess of the protein.

When a ration contains 0.40 of a pound of protein, 0.40 of a pound of fat, and 1 pound of carbohydrates, it has a nutritive ratio of 1 to 4.8 and is a narrow ration.

To calculate the nutritive ratio, the fat is multiplied by 2¼, the product added to the carbohydrates, and this sum divided by the protein. It is not possible to designate accurately the amount of protein and other nutrients that should be in the daily ration of all persons, because the needs of the body vary so with different individuals.

Hard and fast rules governing the amounts of nutrients to be consumed cannot as yet be formulated, as our knowledge of the subject is too limited. It is known that both excessive and scant amounts are alike injurious.

While the appetite may indicate either hunger or satiety, it alone cannot always be relied upon as a safe guide for determining the amount and kind of food to consume, although the demands of appetite should not be disregarded until it has been demonstrated beyond a doubt that it is not voicing the needs of nature.

There has been a tendency which perhaps was a survival of the Puritanical ideas of the early days to stamp as hurtful whatever seemed desirable and pleasant; as examples might be cited the craving for water by fever patients, and for sugar by growing children, which have now been proven to be normal demands of nature.

Dietary Standards: As a result of a large number of dietary studies and digestion experiments, dietary standards have been prepared. Atwater in this country and Voit in Germany have proposed such standards for men employed at different kinds of labour, as follows:

	Protein lb.	*Fat lb.*	*Carbohydrates lb.*	*Fuel Value Calories*	*Nutritive Ratio*
Man with little physical exercise	0.20	0.20	0.66	2450	5.5
Man with light muscular work	0.22	0.22	0.77	2800	5.7
Man with moderate muscular work	0.28	0.28	0.99	3520	5.8
Man with active muscular work	0.33	0.33	1.10	4060	5.6
Man with hard muscular work	0.39	0.55	1.43	5700	6.9

The quantity of nutrients increases with the labour to be performed. In order to secure the necessary heat and energy, rations for men at heavy labour contain proportionally more fat and carbohydrates than are required for light work.

All dietary standards, however, should be regarded as tentative only. Opinions differ greatly on different points; for example, as to the amount of protein a ration should contain. This is a matter that can be determined only from extended investigations under a variety of conditions, and as yet results are too meager to formulate other than tentative standards. Chittenden has found that the body can be sustained on very much less protein than is called for in the standard ration. The amount of protein in the ration should be ample to sustain the body weight and maintain a nitrogen equilibrium; that is, the income and outgo of nitrogen from the body should be practically equal.

Dietaries and Dietary Standards

"While one freely admits that health and a large measure of muscular strength may be maintained upon a minimum supply of protein, yet I think that a dispassionate survey of mankind will show that races which adopt such a diet are lacking in what, for want of a better word, one can only describe as energy."

On the other hand, excessive and unnecessarily large amounts of protein are sometimes consumed, adding greatly to the cost of the ration and necessitating additional labour on the part of the body for its elimination.

Number of Meals Per Day

Some persons advocate two meals per day rather than three, but dietary studies show that the best results are secured when the food is divided among three rather than two meals, and with a two-meal system the tendency is to consume a larger total amount of food than when three meals are eaten. It is not essential that the food be equally divided among the three meals.

Any one of them may be lighter or more substantial as the habits and inclinations of the individual dictate. If it is found necessary to reduce the total quantity of food consumed, this may be done by a proportional reduction of each of the meals, or of any one of them instead of decreasing the number of meals per day. The occasional missing of a meal is sometimes beneficial, in cases of digestion disorders, but the ordinary requirements of persons in normal health who have either mental or physical labour to perform are best met when three meals per day are consumed, as this insures an even supply of nutrients. For persons of sedentary habits, the kind and quantity of food at each meal must be regulated largely by the individual from knowledge based on personal experience.

"In the matter of diet every man must, in the last resort, be a law unto himself; but he should draw up his dietetic code intelligently and apply it honestly, giving due heed to the warnings which nature is sure to address to him should he at any time transgress."

If there is trouble in digesting the food, it is well to study the other habits of life along with the food question, for it may be the difficulty arises from some other cause, and would be remedied by more exercise and fresh air, avoiding rush immediately after meals, more thorough mastication, or less worry. It is a serious matter to shut off the supply of food from a person not suffering from some disease and who is working; as well cut off the supply of fuel from a furnace and then expect a full amount of energy and heat. But unlike the furnace, when the human body is deprived of needed nutrients it preys upon itself and uses up its reserve that should be drawn upon only in cases of illness or extreme nervous strain. Some persons live in such a way as to never have any reserve of strength and energy to call upon but use up each day all the body can produce and so become physical bankrupts when they should be in their prime. Food is required for the production of nerve energy as well as physical energy.

Mixed Dietary Desirable

Experiments in the feeding of farm animals show that the best results come from the combination of a number of foods to form a mixed ration, rather than from the use of one food alone, for in this way the work of digestion is more evenly distributed, and a higher degree of efficiency is secured from the foods consumed. The same is true in human feeding; the best results are secured from a mixed diet. Ordinarily, about two fifths of the nutrients of a ration are derived from animal and three fifths from vegetable sources.

Animal and Vegetable Foods; Economy of Production.—Animal foods can never compete in cheapness of the nutrients with cereals and vegetables, as it takes six to eight pounds or more of a cereal, together with forage crops, to make a pound of meat. Hence the returns in food value are very much larger from the direct use of the cereals as human food, than from the feeding of cereals to cattle and the use of the meat. As the population of a country increases, and foods necessarily become more expensive, cereals are destined to replace animal foods to a great extent, solely as a matter of economy.

Food Habits

Long-established dietary habits and customs are not easily changed, and when the body becomes accustomed to certain foods, substitution of others, although equally valuable, may fail to give satisfactory results. For example, immigrants from southern Europe demand foods with which they are familiar, as macaroni, olive oil, and certain kinds of cheese, foods which are generally imported and more expensive than the staples produced in this country, and when they are compelled to live on other foods, even though they have as many nutrients, they complain of being underfed. Previously acquired food habits appear to affect materially the process of digestion and assimilation. Sudden and pronounced change in the feeding of farm animals is attended with unsatisfactory results, and whenever changes are made in the food of either humans or animals they should be gradual rather than radical.

Underfed Families.—As the purchasing of food is often done by inexperienced persons, palatability rather than nutritive value is made the basis of choice. Dietary studies show that because of lack of knowledge of the nutritive value of foods, whole families are often underfed. Particularly is this true where the means for purchasing foods are limited. In dietary studies among poor families in New York City, the United States Department of Agriculture notes: "It is quite evident that what is needed among these families more than anything else is instruction in the way to make the little they have go the farthest." Some classes of the rich too are equally liable to be underfed, as they are more prone to food notions and are able to indulge them. Among the children of the rich are found some as poorly nourished as among the poor.

Cheap and Expensive Foods

Among the more expensive items of a ration are meats, butter, and canned fruits. The difference in composition and nutritive value between various cuts of meat is small, being largely physical, and affecting taste and flavour rather than nutritive value. Expensive cuts of meat, high-priced breakfast cereals, tropical fruits and foods which impart special flavors, add little in the way of nutritive value to the ration, but greatly enhance the cost of living. Ordinarily the cheapest foods are corn meal, wheat flour and bread, milk, beans, cheese, sugar, and potatoes. The amount of animal and vegetable foods to combine with these to form a balanced ration may be governed largely by personal preference or cost, as there is little difference in nutritive value.

Food Notions

Many erroneous ideas exist as to the nutritive value of foods, and often wholesome and valuable foods are discriminated against because of prejudice. Skim milk is usually regarded as containing little if any nourishing material, when in reality it has a high protein content, and can be added to other foods to increase their nutritive value. The less expensive cuts of meat contain more total nutrients than many of the more expensive ones. Beef extracts have been erroneously said to contain more nutrients than beef, and mushrooms to be equal in value of beefsteak; chemical analyses fail to confirm either statement. The banana also has been overestimated as to food value, and while it contains more nutrients than many fruits, it is not the equal of cereals, as has been claimed. Cocoa, although a valuable beverage, adds but little in the way of nutrients to a ration unless it is made with milk. The value of a food should be based upon its composition as determined by chemical analysis, its digestibility as founded upon digestion experiments, and its palatability and mechanical structure. Food notions have, in many instances, been the cause of banishing from the dietary wholesome and nutritious foods, of greatly increasing the cost of living, as well as of promulgating incorrect ideas in regard to foods, so that individuals and in some cases entire families have suffered from improper or insufficient food.

Dietary of Two Families Compared

A dietary study often reveals ways in which it is possible to improve the ration in kinds and amounts of food, and sometimes at less expense. The following dietaries of two families for the same period show that one family expends over twice as much in the purchase of foods as the other family, and yet the one whose food costs the less actually secures the larger amount of nutritive material and is better fed than the family where more money is expended for food.

THE ORIGIN OF NUTRITIONAL NEEDS

Interactions of the Origins

The nutritional need originates in the chemical reactions that obey the principles of thermodynamics, and arose during the billion years following the formation of the earth. The different energy sources that existed at the time enabled chemical compounds such as water and water vapour (H_2O), methane (CH_4) and ammonia (NH_3) to interact and combine, thus generating the basic components of living matter. Laboratory simulations of the chemical and energy conditions assumed to have prevailed at the beginning of the earth have permitted the recreation of practically all these basic components from these three simple gases. The interaction then continued through the condensation of the basic components in long chains that generated the molecules (polymers) typical of the living world, and later the assembly of polymers in organelles – these in turn formed living cells. It took a billion years for the fi rst single-celled forms of life (or unicellulars) that compose the bacterial kingdom to come into being. It took a further three billion years for unicellulars to develop, and later assemble in, multi-cellular organisms, with the differentiation of organs, and so give rise to the vegetable and animal kingdoms. Finally, the three kingdoms continued to develop during the next 800 million years or so, by fashioning the earth and its atmosphere to bring them approximately to their contemporary form. *Homo sapiens* (or, rather, *Homo economicus*, an expression that better refl ects his exploitation of the environment) appeared only one hundred thousand years ago.

During this evolution, the principles of thermodynamics remained a denominator common to the original energy reactions and the nutritional need.

The universal principles that govern the exchanges of energy.

Thermodynamic Determinism

The evolution from the original simple gases to superior animals occurred in successive stages of association between compatible elements. Each level of association brought a new structure, but also new forms of organization, because interaction takes different forms that are specifi c to each level: atoms do not interact like particles do, nor molecules like atoms, nor polymers like basic molecules, any more than societies interact like their components do. However, the same forces always govern these different forms of interaction that occur when the conditions for exchange of energy are met. In other words, where a receptor can capture a source of energy in compliance with the principle of stability, the result is an energy fl ow. Conditions of interaction obey the principle referred to as thermodynamic determinism that governs everything that can be observed in the universe.

Thermodynamic determinism represents what is common to the original energy interactions and the nutritional need that is under scrutiny here. This nutritional need appeared at the same time as the living cell, the fi rst form of terrestrial life.

The Appearance of the Living Cell

One crucial stage of association was the combination of constitutive elements to form a structure which became a specifi c entity: the living cell. Indeed, the living cell incorporates and specifi cally organizes energy interactions complying with thermodynamic determinism within a material structure limited by a semi-permeable membrane.

This entails several consequences, three of which are of particular interest here.

1. Thermodynamic reactions imply the presence of both a source and a receptor of energy. An organism is alive only through the thermodynamic reactions that characterize its existence; it must therefore have access to a source that can satisfy the need for energy of its material receptors. This source is initially internal, but runs out because the structure of the organism is fi nite in space. Therefore, the source must imperatively be renewed – and permanently so to avoid death – from the environment.
2. Between the original simple entities whose energy exchanges were chaotic, and animals whose energy exchanges occur within the

organism, the energy fl ow was captured in increasingly complex material structures. The latter must reproduce and ensure their continued existence from materials found in the environment.

3. As complex structures developed and living matter diversifi ed into different organisms and evolved towards superior animals, they lost their capacity to create everything from simple entities (a capacity the fi rst cells had, and some unicellulars still have). Some species thus developed a dependency towards the living beings that retained that capacity, of which they must henceforth absorb all or part to extract the nutrients that they cannot create themselves.

This provides the three components of the nutritional need.

1. The renewal of the source of energy that governs reactions.
2. The provision of this source of energy with material support, and an envelope to manage its dissipation.
3. The obtaining of all or part of these material elements from other living species, if the capacity to create them is inadequate or lost.

The vital dependency towards these three components is the essence of the nutritional need. The nutritional need is one of the consequences of the appearance of the cell and the living organism, and it must be viewed in its conceptual framework.

ESSENTIAL DIETARY MINERALS

Some sources state that sixteen dietary minerals are *required* to support human biochemical processes by serving structural and functional roles as well as electrolytes: Sometimes a distinction is drawn between this category and micronutrients.

Protein

A representation of the 3D structure of myoglobin showing coloured alpha helices. This protein was the first to have its structure solved by X-ray crystallography.

Proteins (also known as polypeptides) are organic compounds made of amino acids arranged in a linear chain and folded into a globular form. The amino acids in a polymer are joined together by the peptide bonds between the carboxyl and amino groups of adjacent amino acid residues. The sequence of amino acids in a protein is defined by the sequence of a gene, which is encoded in the genetic code. In general, the genetic code specifies 20 standard amino acids; however, in certain organisms the genetic code can include selenocysteine—and in certain archaea—pyrrolysine. Shortly after or even during synthesis, the residues in a protein are often chemically modified by post-translational modification, which alters the physical and chemical properties, folding, stability, activity, and ultimately, the function of the proteins. Proteins can also work together to achieve a particular function, and they often associate to form stable complexes.

Like other biological macromolecules such as polysaccharides and nucleic acids, proteins are essential parts of organisms and participate in virtually every process within cells. Many proteins are enzymes that catalyze biochemical reactions and are vital to metabolism. Proteins also have structural or mechanical functions, such as actin and myosin in muscle and the proteins in the cytoskeleton, which form a system of scaffolding that maintains cell shape. Other proteins are important in cell signalling, immune responses, cell adhesion, and the cell cycle. Proteins are also necessary in animals' diets, since animals cannot synthesize all the amino acids they need and must obtain essential amino acids from food. Through the process of digestion, animals break down ingested protein into free amino acids that are then used in metabolism. Proteins were first described by the Dutch chemist Gerhardus Johannes Mulder and named by the Swedish chemist Jöns Jakob Berzelius in 1838. The central role of proteins in living organisms was however not fully appreciated until 1926, when James B. Sumner showed that the enzyme urease was a protein. The first protein to be sequenced was insulin, by Frederick Sanger, who won the Nobel Prize for this achievement in 1958. The first protein structures to be solved were hemoglobin and myoglobin, by Max Perutz and Sir John Cowdery Kendrew, respectively, in 1958. The three-dimensional structures of both proteins were first determined by x-ray diffraction analysis; Perutz and Kendrew shared the 1962 Nobel Prize in Chemistry for these discoveries. Proteins may be purified from other cellular components using a variety of techniques such as ultracentrifugation, precipitation, electrophoresis, and chromatography; the advent of genetic engineering has made possible a number of methods to facilitate purification. Methods commonly used to study protein structure and function include immunohistochemistry, site-directed mutagenesis, and mass spectrometry.

Biochemistry

Resonance structures of the peptide bond that links individual amino acids to form a protein polymer.

Most proteins are linear polymers built from series of up to 20 different L-á-amino acids. All amino acids possess common structural features, including an á-carbon to which an amino group, a carboxyl group, and a variable side chain are bonded. Only proline differs from this basic structure as it contains an unusual ring to the N-end amine group, which forces the CO–NH amide moiety into a fixed conformation. The side chains of the standard amino acids, detailed in the list of standard amino acids, have a great variety of chemical structures and properties; it is the combined effect of all of the amino acid side chains in a protein that ultimately determines its three-dimensional structure and its chemical reactivity.

Chemical structure of the peptide bond and a peptide bond between leucine and threonine.

The amino acids in a polypeptide chain are linked by peptide bonds. Once linked in the protein chain, an individual amino acid is called a *residue,* and the linked series of carbon, nitrogen, and oxygen atoms are known as the *main chain* or *protein backbone.* The peptide bond has two resonance forms that contribute some double-bond character and inhibit rotation around its axis, so that the alpha carbons are roughly coplanar. The other two dihedral angles in the peptide bond determine the local shape assumed by the protein backbone. The end of the protein with a free carboxyl group is known as the C-terminus or carboxy terminus, whereas the end with a free amino group is known as the N-terminus or amino terminus.

The words *protein, polypeptide,* and *peptide* are a little ambiguous and can overlap in meaning. *Protein* is generally used to refer to the complete biological molecule in a stable conformation, whereas *peptide* is generally reserved for a short amino acid oligomers often lacking a stable three-dimensional structure. However, the boundary between the two is not well defined and usually lies near 20–30 residues. *Polypeptide* can refer to any single linear chain of amino acids, usually regardless of length, but often implies an absence of a defined conformation.

Synthesis

Proteins are assembled from amino acids using information encoded in genes. Each protein has its own unique amino acid sequence that is specified by the nucleotide sequence of the gene encoding this protein. The genetic code is a set of three-nucleotide sets called codons and each three-nucleotide combination designates an amino acid, for example AUG (adenine-uracil-guanine) is the code for methionine. Because DNA contains four nucleotides, the total number of possible codons is 64; hence, there is some redundancy in the genetic code, with some amino acids specified by more than one codon. Genes encoded in DNA are first transcribed into pre-messenger RNA (mRNA) by proteins such as RNA polymerase. Most organisms then process the pre-mRNA (also known as a *primary transcript*) using various forms of post-transcriptional modification to form the mature mRNA, which is then used as a template for protein synthesis by the ribosome. In prokaryotes the mRNA may either be used as soon as it is produced, or be bound by a ribosome after having moved away from the nucleoid. In contrast, eukaryotes make mRNA in the cell nucleus and then translocate it across the nuclear membrane into the cytoplasm, where protein synthesis then takes place. The rate of protein synthesis is higher in prokaryotes than eukaryotes and can reach up to 20 amino acids per second.

The process of synthesizing a protein from an mRNA template is known as translation. The mRNA is loaded onto the ribosome and is read three nucleotides at a time by matching each codon to its base pairing anticodon located on a transfer RNA molecule, which carries the amino acid corresponding to the codon it recognizes. The enzyme aminoacyl tRNA synthetase "charges" the tRNA molecules with the correct amino acids. The growing polypeptide is often termed the *nascent chain*. Proteins are always biosynthesized from N-terminus to C-terminus.

The size of a synthesized protein can be measured by the number of amino acids it contains and by its total molecular mass, which is normally reported in units of *daltons* (synonymous with atomic mass units), or the derivative unit kilodalton (kDa). Yeast proteins are on average 466 amino acids long and 53 kDa in mass. The largest known proteins are the titins, a component of the muscle sarcomere, with a molecular mass of almost 3,000 kDa and a total length of almost 27,000 amino acids.

Chemical Synthesis

Short proteins can also be synthesized chemically by a family of methods known as peptide synthesis, which rely on organic synthesis techniques such as chemical ligation to produce peptides in high yield. Chemical synthesis allows for the introduction of non-natural amino acids into polypeptide chains, such as attachment of fluorescent probes to amino acid side chains. These methods are useful in laboratory biochemistry and cell biology, though generally not for commercial applications. Chemical synthesis is inefficient for polypeptides longer than about 300 amino acids, and the synthesized proteins may not readily assume their native tertiary structure. Most chemical synthesis methods proceed from C-terminus to N-terminus, opposite the biological reaction.

Structure of Proteins

Three possible representations of the three-dimensional structure of the protein triose phosphate isomerase. Left: all-atom representation coloured by atom type. Middle: Simplified representation illustrating the backbone conformation, coloured by secondary structure. Right: Solvent-accessible surface representation coloured by residue type (acidic residues red, basic residues blue, polar residues green, nonpolar residues white).

Most proteins fold into unique 3-dimensional structures. The shape into which a protein naturally folds is known as its native conformation. Although many proteins can fold unassisted, simply through the chemical properties of their amino acids, others require the aid of molecular chaperones to fold into their native states. Biochemists often refer to four distinct aspects of a protein's structure:

Primary Structure: The Amino Acid Sequence

Secondary Structure: regularly repeating local structures stabilized by hydrogen bonds. The most common examples are the alpha helix, beta sheet and turns. Because secondary structures are local, many regions of different secondary structure can be present in the same protein molecule.

Tertiary Structure: the overall shape of a single protein molecule; the spatial relationship of the secondary structures to one another. Tertiary structure is generally stabilized by nonlocal interactions, most commonly the formation of a hydrophobic core, but also through salt bridges, hydrogen bonds, disulfide bonds, and even post-translational modifications. The term "tertiary structure" is often used as synonymous with the term *fold*. The tertiary structure is what controls the basic function of the protein.

Quaternary Structure: the structure formed by several protein molecules (polypeptide chains), usually called *protein subunits* in this context, which function as a single protein complex.

Proteins are not entirely rigid molecules. In addition to these levels of structure, proteins may shift between several related structures while they perform their functions. In the context of these functional rearrangements, these tertiary or quaternary structures are usually referred to as "conformations", and transitions between them are called *conformational changes.* Such changes are often induced by the binding of a substrate molecule to an enzyme's active site, or the physical region of the protein that participates in chemical catalysis. In solution proteins also undergo variation in structure through thermal vibration and the collision with other molecules.

Proteins can be informally divided into three main classes, which correlate with typical tertiary structures: globular proteins, fibrous proteins, and membrane proteins. Almost all globular proteins are soluble and many are enzymes. Fibrous proteins are often structural, such as collagen, the major component of connective tissue, or keratin, the protein component of hair and nails. Membrane proteins often serve as receptors or provide channels for polar or charged molecules to pass through the cell membrane.

A special case of intramolecular hydrogen bonds within proteins, poorly shielded from water attack and hence promoting their own dehydration, are called dehydrons.

Structure Determination

Discovering the tertiary structure of a protein, or the quaternary structure of its complexes, can provide important clues about how the protein performs its function. Common experimental methods of structure determination include X-ray crystallography and NMR spectroscopy, both of which can produce information at atomic resolution. Dual polarisation interferometry is a quantitative analytical method for measuring the overall protein conformation and conformational changes due to interactions or other stimulus.

Circular dichroism is another laboratory technique for determining internal beta sheet/helical composition of proteins. Cryoelectron microscopy is used to produce lower-resolution structural information about very large protein complexes, including assembled viruses; a variant known as electron crystallography can also produce high-resolution information in some cases, especially for two-dimensional crystals of membrane proteins. Solved structures are usually deposited in the Protein Data Bank (PDB), a freely available resource from which structural data about thousands of proteins can be obtained in the form of Cartesian coordinates for each atom in the protein.

Many more gene sequences are known than protein structures. Further, the set of solved structures is biased toward proteins that can be easily subjected to the conditions required in X-ray crystallography, one of the major structure determination methods. In particular, globular proteins are comparatively easy to crystallize in preparation for X-ray crystallography. Membrane proteins, by contrast, are difficult to crystallize and are underrepresented in the PDB. Structural genomics initiatives have attempted to remedy these deficiencies by systematically solving representative structures of major fold classes. Protein structure prediction methods attempt to provide a means of generating a plausible structure for proteins whose structures have not been experimentally determined.

IN NUTRITION AND DISEASES

Vitamins are essential for the normal growth and development of a multicellular organism. Using the genetic blueprint inherited from its parents, a fetus begins to develop, at the moment of conception, from the nutrients it absorbs. It requires certain vitamins and minerals to be present at certain times. These nutrients facilitate the chemical reactions that produce among other things, skin, bone, and muscle. If there is serious deficiency in one or more of these nutrients, a child may develop a deficiency disease. Even minor deficiencies may cause permanent damage.

For the most part, vitamins are obtained with food, but a few are obtained by other means. For example, microorganisms in the intestine—commonly known as "gut flora"—produce vitamin K and biotin, while one form of vitamin D is synthesized in the skin with the help of the natural ultraviolet wavelength of sunlight.

Humans can produce some vitamins from precursors they consume. Examples include vitamin A, produced from beta carotene, and niacin, from the amino acid tryptophan.

Once growth and development are completed, vitamins remain essential nutrients for the healthy maintenance of the cells, tissues, and organs that make up a multicellular organism; they also enable a multicellular life form to efficiently use chemical energy provided by food it eats, and to help process the proteins, carbohydrates, and fats required for respiration.

Deficiencies

Humans must consume vitamins periodically but with differing schedules, to avoid deficiency. Human bodily stores for different vitamins vary widely; vitamins A, D, and B_{12} are stored in significant amounts in the human body, mainly in the liver, and an adult human's diet may be deficient in vitamins A for many months and B_{12} in some cases for years, before developing a deficiency condition.

However, vitamin B_3 (niacin and niacinamide) is not stored in the human body in significant amounts, so stores may only last a couple of weeks.. For vitamin C, the first symptoms of scurvy in experimental studies of complete vitamin C deprivation in humans have varied widely, from a month to more than six months, depending on previous dietary history which determined body stores. No vitamin must be consumed every day, however, or even every week.

Deficiencies of vitamins are classified as either primary or secondary. A primary deficiency occurs when an organism does not get enough of the vitamin in its food. A secondary deficiency may be due to an underlying disorder that prevents or limits the absorption or use of the vitamin, due to a "lifestyle factor", such as smoking, excessive alcohol consumption, or the use of medications that interfere with the absorption or use of the vitamin. People who eat a varied diet are unlikely to develop a severe primary vitamin deficiency. In contrast, restrictive diets have the potential to cause prolonged vitamin deficits, which may result in often painful and potentially deadly diseases.

Well-known human vitamin deficiencies involve thiamine (beriberi), niacin (pellagra), vitamin C (scurvy) and vitamin D (rickets). In much of the developed world, such deficiencies are rare; this is due to (1) an adequate supply of food; and (2) the addition of vitamins and minerals to common foods, often called fortification. In addition to these classical vitamin deficiency diseases, some evidence has also suggested links between vitamin deficiency and a number of different disorders.

Side Effects and Overdose

In large doses, some vitamins have documented side effects that tend to be more severe with a larger dosage. The likelihood of consuming too much of any vitamin from food is remote, but overdosing (vitamin poisoning) from vitamin supplementation does occur. At high enough dosages some vitamins cause side effects such as nausea, diarrhea, and vomiting.

When side effects emerge, recovery is often accomplished by reducing the dosage. The doses of vitamins different individual can tolerate varies widely, and appear to be related to age and state of health.

In the United States, overdose exposure to all formulations of "vitamins" was reported by 62,562 individuals in 2004 (nearly 80% of these exposures were in children under the age of 6), leading to 53 "major" life-threatening outcomes and 2 deaths;. This may be compared to the 19,250 people who died of unintentional poisoning of all kinds in the U.S. in the same year (2004). In 2007, 58,000 exposures to various vitamins and multivitamin-mineral formulations were reported to poison control centers, which resulted in 17 severe reactions and 1 death. In 2008 there were no deaths reported from vitamin overdose but represented 2.8% of all exposures reported to poison control and 68,911 individual cases.

Supplements

Dietary supplements, often containing vitamins, are used to ensure that adequate amounts of nutrients are obtained on a daily basis, if optimal amounts of the nutrients cannot be obtained through a varied diet. Scientific evidence supporting the benefits of some vitamin supplements is well established for certain health conditions, but others need further study. In some cases, vitamin supplements may have unwanted effects, especially if taken before surgery, with other dietary supplements or medicines, or if the person taking them has certain health conditions. Dietary supplements may also contain levels of vitamins many times higher, and in different forms, than one may ingest through food.

There have been mixed studies on the importance and safety of dietary supplementation. One study released in May 2009 found that antioxidants such as vitamins C and E may actually curb some benefits of exercise. A meta-analysis published in 2006 suggested that Vitamin A and E supplements not only provide no tangible health benefits for generally healthy individuals, but may actually increase mortality, although two large studies included in the analysis involved smokers, for which it was already known that beta-carotene supplements can be harmful. National Institutes of Health (NIH) suggests that developments in vitamin E research show that the studies are flawed, and that "evidence for toxicity of a specific form of tocopherol in excess may not be used to conclude that high-dosage "vitamin E" supplementation may increase all-cause mortality." Moreover, NIH states that "the diets of most Americans provide less than the RDA levels of vitamin E." The majority of the medical community appears to agree with popular consensus that dietary supplementation is important. In a peer reviewed study, Council for Responsible Nutrition found that "72 percent of physicians and 89 percent of nurses used dietary supplements and that 79 percent of physicians and 82 percent of nurses said that they recommend dietary supplements to their patients." In spite of contradictory studies, many consumers believe they don't get enough nutrients in their food supply, and the importance of supplementation is evidenced by the worldwide multi-billion dollar revenue generated by the health supplement industry.

COST AND NUTRIENT CONTENT OF FOODS

The market price and the nutritive value of foods are often at variance, as those which cost the most frequently contain the least nutrients. It is difficult to make absolute comparisons as to the nutritive value of foods at different prices, because they differ not only in the amounts, but also in the kinds of nutrients. While it is not possible to express definitely the value of one food in terms of another, approximate comparisons may be made as to the amounts of nutrients that can be secured for a given sum of money when foods are at different prices, and tables have been prepared making such comparisons.

Composition of Foods

Nutrients Procurable for a Given Sum: To ascertain the nutrients procurable for a given sum first determine the amount in pounds that can be obtained, say, for ten cents, and then multiply by the percentages of fat, protein, carbohydrates, and calories in the food. The results are the amounts, in pounds, of nutrients procurable for that sum of money. For example: if milk is 5 cents per quart, two quarts or approximately four pounds, can be procured for 10 cents. If the milk contains fat, 4 per cent, protein, 3.3 per cent, carbohydrates, 5 per cent, and fuel value, 310 calories per pound, multiplying each of these by 4 gives the nutrients and fuel value in four pounds, or 10 cents worth of milk, as follows:

Protein	0.13 lb.
Fat	0.16 lb.
Carbohydrates	0.2 lb.
Calories	1240

If it is desired to compare milk at 5 cents per quart with round steak at 15 cents per pound, 10 cents will procure 0.66, or two thirds of a pound of round steak containing on an average (edible portion) 19 per cent protein, 12.8 per cent fat, and yielding 890 calories per pound. If 10 per cent is refuse, there is edible about 0.6 of a pound. The amounts of nutrients in the 0.6 of a pound of steak, edible portion, or 0.66 lb. as purchased would be:

Protein	0.11 lb.
Fat	0.08 lb.
Calories	534

It is to be observed that from the 10 cents' worth of milk a little more protein, 0.08 of a pound more fat, and nearly two and one half times as many calories can be secured as from the 10 cents' worth of meat. This is due to the carbohydrates and the larger amount of fat which the milk contains. At these prices, milk should be used liberally in the dietary, as it furnishes more of all the nutrients than does meat.

It would not be advisable to exclude meat entirely from the ration, but milk at 5 cents per quart is cheaper food than meat at 15 cents per pound. In making comparisons, preference cannot always be given to one food because of its containing more of any particular nutrient, for often there are other factors that influence the value.

Comparing Foods as to Nutritive Value

In general, preference should be given to foods which supply the most protein, provided the differences between the carbohydrates and fats are not large.

When the protein content of two foods is nearly the same, but the fats and carbohydrates differ materially, the preference may safely be given to the food which supplies the larger amount of total nutrients.

A pound of protein in a ration is more valuable than a pound of either fat or carbohydrates, although it is not possible to establish an absolute scale as to the comparative value of these nutrients, because they serve different functional purposes in the body. It is sometimes necessary to use small amounts of foods rich in protein in order to secure a balanced ration; excessive use of protein, however, is not economical, as that which is not needed for functional purposes is converted into heat and energy which could be supplied as well by the carbohydrates, and they are less expensive nutrients.

PECUNIARY ECONOMY OF FOOD

Ten Cents will Purchase

It is to be noted in the table that, ordinarily, for the same amount of money the most nutrients can be obtained in the form of milk, cheese, sugar, and beans, corn meal, wheat flour, oatmeal, and cereals in bulk.

While meats supply protein liberally, they fail to furnish carbohydrates as the vegetables.

Dietary Studies of Families, unnecessarily expensive foods are often used, resulting either in lack of nutrients or unbalanced rations.

Examples:

1. Compute the calories and the amounts of protein, fat, and carbohydrates that can be procured for 25 cents in cheese selling for 18 cents per pound; how do these compare with the nutrients in eggs at 20 cents per dozen?
2. Which food furnishes the larger amount of nutrients, potatoes at 50 cents per bushel or flour at $6 per barrel?
3. How do beans at 10 cents per quart compare in nutritive value with beef at 15 Cents per pound?
4. How does salt codfish at 10 cents per pound compare in nutritive value with lamb chops at 15 cents per pound?
5. Compare in nutritive value cream at 25 cents per quart with butter at 30 cents per pound.
6. Calculate the composition and nutritive value of a cake made of sugar, 8 oz.; butter, 4 oz.; eggs, 8 oz.; flour, 8 oz.; and milk, 4 oz.; the baked cake weighs one and three fourths pounds.

THE PHENOMENON OF ORGANIZED ASSOCIATION

The nutritional need is the effective and direct consequence of the confi nement of molecular interactions to a specifi c spatial structure – that is, the cell or living organism. More generally, from a conceptual perspective, it results from the phenomenon of organized association.

Its principles are as follows.

1. The association occurs within a defi ned structure in which the elements of the association perform activities.
2. This structure makes sense only because activities are organized within it; for this, they must meet four conditions:
   - anchorage on a material support that provides them with a hold; q compliance with a norm;

- subjection to a system of control that ensures the maintenance of, and respect for, the norm; q ability to be replicated, through the transmission to successors of the elements of the group of everything required to perform the activities according to the norms specifi c to the organization. These four conditions represent the basis of organized behaviour.

3. Organized association and behaviour imply specifi c demands that arise from the four conditions discussed above. These demands are as follows: q provision of the material support required to implement the activities; q provision of a norm; q provision of a control system; q capability to replicate the association.
4. Organized association thus entails obligations that manifest themselves on two separate levels: q on the algorithmic level of any organized behaviour, activities must meet the four conditions stated under point 2 above, which constitute the determinism of the activities; q on the level specifi c to the association under consideration, the demands stated under point 3 above must be satisfi ed – these demands defi ne its needs, which in turn represent the determinism of the association.
5. The conditions of organized behaviour, together with the needs that result from it, constitute the determinism imposed upon any organized group when it implements its activities. It is also worth noting that the determinism specifi c to the association includes the needs of the elements that compose it, and each new level of needs is related to the previous level. Generally, the global determinism (i.e. algorithmic and specifi c) of an association is defi ned according to the level of organization under consideration: thermodynamic in the case of particles, atoms or molecules, biological in the case of living organisms, cultural in the case of humans in a group.

The nutritional need is a direct result of determinism insofar as the cell or living organism constitutes organized associations.

Indeed, if the living organism is an entity whose behaviour is organized, then it follows that the activities that occur within it rest on a specifi c basis, that they obey the laws of thermodynamics, and that they are subjected to the balance between substrate and product or by a neuro-endocrine system.

It also follows that the organism produces replacements for its elements by using a genetic code that permits the duplication of all the elements in such a way as to function like their predecessors. For the living organism, the same demands exist, related to the following:

- the material support of the structure – the three components of nutritional need;

- the application of the norm – the laws of metabolism;
- the control of the application – among others, the necessary neuro-endocrine system among animals;
- the specifi c replication of the constituents – the genetic code, present in any living cell.

The fi nal behaviour (i.e. thermodynamic interactions and compliance with the conditions for existence) will result from the biological determinism of the organism under consideration.

2

Nutrition

INTRODUCTION TO NUTRITION

For every physical activity, the body requires energy and the amount depends on the duration and type of activity. Energy is measured in Calories and is obtained from the body stores or the food we eat. Glycogen is the main source of fuel used by the muscles to enable you to undertake both aerobic and anaerobic exercise. If you train with low glycogen stores, you will feel constantly tired, training performance will be lower and you will be more prone to injury and illness. A calorie (cal) is the amount of heat energy required to raise the temperature of 1g of water 1°C from 14° to 15°C. A kilocalorie (kcal) is the amount of heat required to raise the temperature of 1000g of water 1°C.

Nutrient Balance

Carefully planned nutrition must provide an energy balance and a nutrient balance.

The nutrients are:

- Proteins - essential to growth and repair of muscle and other body tissues
- Fats - one source of energy and important in relation to fat soluble vitamins
- Carbohydrates - our main source of energy
- Minerals - those inorganic elements occurring in the body and which are critical to its normal functions
- Vitamins - water and fat soluble vitamins play important roles in many chemical processes in the body
- Water - essential to normal body function - as a vehicle for carrying other nutrients and because 60% of the human body is water

- Roughage - the fibrous indigestible portion of our diet essential to health of thedigestive system

What are the daily energy requirements?

Personal energy requirement = basic energy requirements + extra energy requirements

Basic energy requirements (BER) includes your basal metabolic rate (BMR) and general daily activities

- For every Kg of body weight 1.3 Calories is required every hour. (An athlete weighing 50Kg would require 1.3 × 24hrs × 50Kg = 1560 Calories/ day)
- For a calculation of your BMR, see the calculator on the Resting Daily Energy Expenditure (RDEE) page

Extra energy requirements (EER)

- For each hours training you require an additional 8.5 Calories for each Kg of body weight. (For a two hour training session our 50Kg athlete would require 8.5 × 2hrs × 50Kg = 850 Calories)

An athlete weighing 50Kg who trains for two hours would require an intake of approximately 2410 Calories (BER + EER = 1560 + 850)

Energy Fuel

Like fuel for a car, the energy we need has to be blended. The blend that we require is as follows:

- 57% Carbohydrates (sugar, sweets, bread, cakes)
- 30% Fats (dairy products, oil)
- 13% Protein (eggs, milk, meat, poultry, fish)

The energy yield per gram is as follows: Carbohydrate - 4 Calories, Fats - 9 Calories and Protein - 4 Calories.

What does a 50 kg athlete require in terms of carbohydrates, fats and protein?

- Carbohydrates - 57% of 2410 = 1374 Calories - at 4 Calories/gram = 1374 ÷ 4 = 343 grams
- Fats - 30% of 2410 = 723 Calories - at 9 Calories/gram = 723 ÷ 9 = 80 grams
- Protein - 13% of 2410 = 313 Calories - at 4 Calories/gram = 313 ÷ 4 = 78 grams

What types of fat are there?

The nature of the fat depends on the type of fatty acids that make up the triglycerides. All fats contain both saturated and unsaturated fatty acids but are usually described as 'saturated' or 'unsaturated' according to the proportion of fatty acids present. Saturated fats are generally solid at room temperature and tend to be animal fats. Unsaturated fats are liquid at room temperature and are usually vegetable fats - there are exceptions e.g. palm oil, a vegetable oil that contains a high percentage of saturated fatty acids.

Unsaturated	Saturated
Sunflower oil	Beef
Olive Oil	Bacon
Rice Oil	Cheese
Nuts	Butter
Rapeseed Oil	Biscuits
Oily fish - Sardines	Crisps

What types of carbohydrates are there?

There are two types of carbohydrates - starchy (complex) carbohydrates and simple sugars. The simple sugar's are found in confectionery, muesli bars, cakes and biscuits, cereals, puddings, soft drinks and juices and jam and honey but they also contain fat. Starchy carbohydrates are found in potatoes, rice, bread, wholegrain cereals, semi skimmed milk, yoghurt, fruit, vegetables, beans and pulses. Both types effectively replace muscle glycogen. The starchy carbohydrates are the ones that have all the vitamins and minerals in them as well as protein. They are also low in fat as long as you do not slap on loads of butter and fatty sauces. The starchy foods are much more bulky so there can be a problem in actually eating that amount of food so supplementing with simple sugar alternatives is necessary. Your digestive system converts the carbohydrates in food into glucose, a form of sugar carried in the blood and transported to cells for energy. The glucose, in turn, is broken down into carbon dioxide and water. Any glucose not used by the cells is converted into glycogen - another form of carbohydrate that is stored in the muscles and liver. However, the body's glycogen capacity is limited to about 350 grams; once this maximum has been reached, any excess glucose is quickly converted into fat. Base your main meal with the bulk on your plate filled with carbohydrates and small amounts of protein such as meat, poultry and fish. The extra protein & vitamins you may require will be in the starchy carbohydrates.

Lactose Intolerance

Lactose intolerance results when the mucosal cells of the small intestine fail to produce lactase that is essential for the digestion of lactose. Symptoms include diarrhoea, bloating, and abdominal cramps following consumption of milk or dairy products.

Carbohydrates for Performance

To support a training session or competition athletes need to eat at an appropriate time so that all the food has been absorbed and their glycogen stores are fully replenished.

Following training & competition, an athlete's glycogen stores are depleted. In order to replenish them the athlete needs to consider the speed at which carbohydrate is converted into blood glucose and transported to the muscles. The rapid replenishment of glycogen stores is important for the track athlete who has a number of races in a meeting.

The rise in blood glucose levels is indicated by a food's Glycaemic Index (GI) - the faster and higher the blood glucose rises the higher the GI.

High GI foods take 1 to 2 hours to be absorbed and low GI foods can take 3 to 4 hours to be absorbed.

Studies have shown that consuming high GI carbohydrates (approximately 1grm per kg body) within 2 hours after exercise speeds up the replenishment of glycogen stores and therefore speeds up recovery time.

Glycogen stores will last for approximately 10 to 12 hours when at rest (sleeping) so this is why breakfast is essential.

Eating 5-6 meals or snacks a day, will help maximise glycogen stores and energy levels, minimise fat storage and stabilise blood glucose and insulin levels.

Eating and Competition

What you eat on a day-to-day basis is extremely important for training. Your diet will affect how fast and how well you progress, and how soon you reach competitive standard.

The page on Nutritional Tips provides some general nutritional advice to help you manage your weight and body fat.

Once you are ready to compete, you will have a new concern: your competition diet. Is it important? What should you eat before your competition? When is the best time to eat? How much should you eat? Should you be eating during the event? In addition, what can you eat between heats or matches? A lot of research has been done in this area, and it is clear that certain dietary approaches can enhance competition performance.

Body Fuel - food for sport

Body Fuel - Food For Sport dissects the major current debates in sports nutrition, analyses the very latest scientific findings - then spells out in plain English their significance for the serious athlete. Every page of this brand new report draws on the latest evidence-based thinking in sports science research - new findings that probably won't percolate through to the general sporting press for many, many months, if they make it at all.

What do I need to do?

Calculate your daily basic and extra requirements, monitor your daily intake (especially your carbohydrates) and then adjust your diet to meet your daily requirements. A good balanced diet should provide you with the required nutrients but does needs to be monitored. The simplest way to monitor the 'energy balance' is to keep a regular check of your weight.

Key factors in your training diet

Each day have three main meals and two to three snacks. All meals should contain both carbohydrate and protein - 20 to 30 grams worth of protein with each main meal and 10 to 20 grams with each snack.

The amount of carbohydrate will vary greatly, mainly depending on your workload. It may be in the region of 40 to 60 grams for main meals and 20 to 30 grams for snacks. If you are training hard and possibly doing multiple daily sessions, the recovery meal is critical.

Have 1grm of carbohydrate per kg of body weight and about 30 grams of protein. Have a drink (e.g. a recovery drink or a pint of skimmed milk) and a banana immediately post-training (this provides about 10 grams of protein and 30 grams of carbohydrate) followed within about 45 minutes with more substantial food such as beans on toast and tuna. Always try to eat at least five pieces of fruit per day. Skimmed milk is a great protein food and provides critical minerals, such as calcium and phosphorous.

Food Composition Tables

Food composition tables are widely used to assess nutrient and energy intakes, and to plan meals. The composition of food can vary widely, depending, among other factors, on the variety of plant or animal, on growing and feeding conditions and, for some foods, on freshness. Tables are based on average values from a number of samples analysed in the laboratory and therefore only provide a rough guide.

Free Calculator

- Calorie Calculator - a free Microsoft Excel spreadsheet that you can download and use on your computer. The spreadsheet will be loaded into a new window.

NUTRIENT REQUIREMENTS DURING LACTATION

Energy

An additional 550 kcal for first six months and 400 kcal during the next six months are required for a lactating mother. This can be achieved by including whole grain cereals, pulses, milk and milk products in the daily diet. There is no need for high calorie foods like ghee.

Protein

Protein requirement is also increased during lactation due to the milk production. The Indian council of medical research has recommended an additional daily intake of 25gm for the first six months and 18 gm during the six to twelve months of lactation. If the protein and energy is lacking in the mothers diet there will be reduction in milk volume rather than in milk quality but if very low protein are taken then the proportion of casein may be reduced. The extra protein can be achieved by including protein rich foods like milk and milk products, egg, meat, fish, poultry, nuts, and cereal and pulse combinations.

Calcium

Calcium is one of the important minerals required both during pregnancy and during lactation. During pregnancy it is required for the foetal skeletal development and during lactation for the milk production. The first three months of the baby is fed only on the breast milk so again the baby depends on the mother for the nutrient requirement. An additional 1000mg per day calcium is required during this period.500 ml milk or milk products per day must be taken for the calcium requirement and the remaining from the other food stuff like ragi, Bengal gram whole, soyabean, amaranth, fenugreek leaves and fish.

Iron

There is no additional requirement of iron during lactation. The baby is born with adequate reserves of iron since milk is not a good source of iron. The additional iron in the mother's diet during lactation does not pass on additional iron to the infant but iron rich foods are essential for mother's own health and the requirement is same as the adult women.

Vitamin

As calories, proteins, and minerals are increased during lactation vitamin requirement also stand increased. Vitamin A, B1, B2, C, niacin, folic acid and vitamin B12 are the major vitamin requirement increased during lactation.

Water

Apart from all these, role of water is very important in the diet for a lactating mother. Fluid is necessary for adequate milk production. Major content of milk is water. Drink 10 - 12 glasses of water every day.

Points to be remembered during lactation" src

1. The diet should include plenty of fresh fruit, vegetables, milk and milk products, fish, poultry, nuts, whole grains, parboiled rice and pulses to provide all extra vitamins, minerals, protein and energy required during this period
2. The number of meals can be increased to meet all the nutritional needs.
3. The food items that help to stimulate the milk production are milk, garlic, fenugreek seeds, moong dhal and cumin. Some believe consuming goat's meat, liver and fish can increase breast milk secretion. There is not much scientific proof available on this.
4. Weight gain should be monitored. The time when baby is switched over to other food, mother should reduce her calorie intake. Other wise chances of excessive weight gain are more.
5. Avoid all emotional problems, and anxiety.
6. Have confidence in your self that you can lactate enough milk for your baby. Remember lactation is controlled by hormone produced by a gland situated near your brain.
7. All most all drugs reach babies through breast milk so before taking any medicine consult a doctor.
8. Alcohol and smoking should be avoided since it can reach baby through breast milk and make baby feel sleepy and make nervous and irritable.
9. Be baby friendly and start breast-feeding soon after delivery.

PREGNANCY AND LACTATION WITH DIABETES

The goals of medical nutrition therapy for pregnancy are to provide adequate maternal and fetal nutrition, energy intake for appropriate weight gain, and any necessary vitamin and mineral supplements. During pregnancy complicated by diabetes, medical nutrition therapy is also important in achieving and maintaining optimal glycemic control.

Unless a woman begins pregnancy with depleted body reserves, energy needs do not increase in the first trimester. An additional 300 kcal/day are suggested during the second and third trimester for increases in maternal blood volume and increases in breast, uterus and adipose tissue, placental growth, fetal growth, and amniotic fluids. However, successful pregnancy outcomes have been reported with lower energy intakes.

In addition to adequate energy intake, pregnant women need to eat a healthy diet with adequate protein (0.75 g $\cdot$ kg^{-1} $\cdot$ day^{-1} plus an additional 10 g/day). Nutrient requirement during pregnancy and lactation are similar for women with and without diabetes. For all women who are capable of becoming pregnant, 400 ìg/day of folic acid from fortified foods and/or a supplement, as well as food folate from a variety of foods, is recommended for the prevention of neural tube defects and other congenital abnormalities. During pregnancy, a healthy diet resulting in appropriate weight gain generally supplies all vitamins and minerals needed. There is inadequate evidence to support prenatal vitamin-mineral supplementation; however, they are often prescribed because of uncertainty of nutrition status and intake. Assessment of the pregnant woman's eating patterns may yield specific individual needs.

The Food and Drug Administration has approved four non-nutritive sweeteners, which are safe to use during pregnancy. As with nondiabetic women, women with diabetes should avoid alcoholic beverages during pregnancy.

Pregnancy with prior-onset type 1 or type 2 diabetes

Prepregnancy nutrition therapy includes an individual prenatal food/ meal plan to optimize blood glucose control. During pregnancy, the distribution of the energy intake and carbohydrates in the meal plan should be based on the woman's food and eating habits, blood glucose records, and the expected physiological effects of pregnancy on her body. Regular meals and snacks are important to avoid hypoglycemia due to the continuous fetal draw of glucose from the mother. An evening snack is usually necessary to decrease the potential for overnight hypoglycemia and fasting ketosis. Blood glucose monitoring and daily food records provide valuable information for insulin and meal plan adjustments.

Gestational diabetes mellitus

Nutrition therapy for gestational diabetes promotes nutrition for maternal and fetal health with adequate energy levels for appropriate gestational weight gain, achievement and maintenance of normoglycemia, and absence of ketones. Carbohydrate is distributed throughout the day into three small-to-moderate sized meals and 2–4 snacks. An evening snack may be needed to prevent accelerated ketosis overnight. Carbohydrate is generally less well tolerated at breakfast than at other meals. Specific nutrition and food recommendations are determined and modified based on individual assessment and self-blood glucose monitoring data.

Energy intake should provide for a desirable weight gain during pregnancy. Hypocaloric diets in obese women with gestational diabetes result in ketonemia and ketonuria. A more modest energy restriction (30% calorie-restriction of estimated energy needs) appears to reduce mean blood glucose levels without elevation in plasma free fatty acids or ketonuria. Daily food records, weekly weight checks, and ketone testing can be used to determine individual energy recommendations and whether a woman is undereating to avoid insulin therapy. Regular aerobic exercise has been shown to lower fasting and postprandial glucose concentrations and may be used as an adjunct to nutrition therapy to improve maternal glycemia. There is insufficient evidence to recommend any specific type of exercise. Blood glucose data are necessary to evaluate the effectiveness of nutrition therapy, exercise, and the need for pharmacological therapy.

If insulin therapy is added to nutrition therapy, maintaining carbohydrate consistency at meals and snacks to facilitate insulin adjustments becomes a primary goal.

Although most women with gestational diabetes revert to normal glucose tolerance postpartum, they are at increased risk of developing gestational diabetes in subsequent pregnancies and type 2 diabetes later in life. Lifestyle modifications aimed at reducing or preventing weight gain and increasing physical activity after pregnancy is recommended and can reduce the risk of subsequent diabetes.

Lactation

Breastfeeding is recommended for women with pre-existing diabetes or gestational diabetes; however, successful lactation requires planning and coordination of care. Breastfeeding lowers blood glucose, often requiring insulin-treated women to eat a snack containing carbohydrate either before or during breastfeeding. Energy requirements during the first 6 months of lactation require an additional <“200 calories above the pregnancy meal plan. However, an energy intake of <“1,800 kcal/day usually meets the nutritional requirements for lactation and may allow for a gradual weight loss.

Expert opinion

- Nutrition requirements during pregnancy and lactation are similar for women with and without diabetes.
- Medical nutrition therapy for gestational diabetes focuses on food choices for appropriate weight gain, normoglycemia, and absence of ketones.
- For some women with gestational diabetes, modest energy and carbohydrate restriction may be appropriate.

OLDER ADULTS WITH DIABETES

There is limited research on changing nutritional needs with aging and virtually none in subjects with diabetes. Therefore, nutrition recommendations for older adults with diabetes must be extrapolated from what is known from the general population. The most reliable indicator of poor nutritional status in the elderly is probably a change in body weight. In general, involuntary gain or loss of >10 pounds or 10% body weight in <6 months indicates a need to evaluate if the reason is nutrition-related.

The need for weight loss in overweight older adults should be carefully evaluated. Older people with diabetes, especially those in long-term care facilities, tend to be underweight rather than overweight. Low body weight has been associated with greater morbidity and mortality in this age group.

Exercise training can significantly reduce the decline in maximal aerobic capacity (VO_2) that occurs with age, improve risk factors for atherosclerosis, slow the decline in age-related lean body mass, decrease central adiposity, and improve insulin sensitivity; all of which is beneficial for the older adult with diabetes.

A daily multivitamin supplement may be appropriate for older adults, especially those with reduced energy intake. All older adults should be advised to have a calcium intake of at least 1,200 mg daily.

The imposition of dietary restrictions on elderly residents with diabetes in long-term health facilities is not warranted. Malnutrition and dehydration may develop because of lack of food choices, poor quality of food, and unnecessary restrictions. Specialized diabetic diets do not appear to be superior to standard (regular) diets in such settings. Therefore, it is recommended that residents are served the regular (unrestricted) menu with consistency in the amount and timing of carbohydrate. There is no evidence to support diets such as "no concentrated sweets" or "no sugar added," which are often served to the elderly in long-term care facilities. Furthermore, it may often be preferable to make medication changes to control blood glucose than to implement food restrictions.

A-Level evidence

- Energy requirements for older adults are less than for younger adults.
- Physical activity should be encouraged.

Expert consensus

- In the elderly, undernutrition is more likely than overnutrition, and therefore caution should be exercised when prescribing weight loss diets.

INFANT

An infant (from the Latin word *infans,* meaning "unable to speak" or "speechless") is the very young offspring of a human or other mammal. When applied to humans, the term is usually considered synonymous with baby, but the latter is commonly applied to the young of any animal. When a human child learns to walk, the term *toddler* may be used instead. The term *infant* is typically applied to young children between the ages of 1 month and 12 months; however, definitions vary between birth and 3 years of age. A newborn is an infant who is only hours, days, or up to a few weeks old. In medical contexts, newborn or neonate(from Latin, *neonatus*, newborn) refers to an infant in the first 28 days after birth; the term applies to premature infants, postmature infants, and full term infants. Before birth, the term*fetus* is used.

Infant is also a legal term referring to any child under the age of legal adulthood.

Physical characteristics

A newborn's shoulders and hips are wide, the abdomen protrudes slightly, and the arms and legs are relatively long with respect to the rest of their body. In first world nations, the average birth weight of a full-term newborn is approximately 3.2 kg.(7 ½ lbs), and is typically in the range of 2.7–4.6 kg (5.5–10 pounds). The average total body length is 35.6–50.8 cm (14–20 inches), although premature newborns may be much smaller. TheApgar score is a measure of a newborn's transition from the uterus during the first minutes after birth.

A newborn's head is very large in proportion to the body, and the cranium is enormous relative to his or her face. While the adult human skull is about 1/8 of the total body length, the newborn's is about 1/4. At birth, many regions of the newborn's skull have not yet been converted to bone, leaving "soft spots" known as fontanels. The two largest are the diamond-shaped anterior fontanel, located at the top front portion of the head, and the smaller triangular-shaped posterior fontanel, which lies at the back of the head. Later in the child's life, these bones will fuse together in a natural process. A protein called noggin is responsible for the delay in an infant's skull fusion.

During labour and birth, the infant's skull changes shape to fit through the birth canal, sometimes causing the child to be born with a misshapen or elongated head. It will usually return to normal on its own within a few days or weeks. Special exercises sometimes advised by physicians may assist the process.

Some newborns have a fine, downy body hair called lanugo. It may be particularly noticeable on the back, shoulders, forehead, ears and face of premature infants. Lanugo disappears within a few weeks. Infants may be born with full heads of hair; others, particularly white infants, may have very fine hair or may even be bald. Amongst fair-skinned parents, this fine hair may be blond, even if the parents are not. The scalp may also be temporarily bruised or swollen, especially in hairless newborns, and the area around the eyes may be puffy.

Newborns' digestive tracts, which of course have never been used prior to birth, are filled with a greenish-black, sticky material called meconium.

This has the function of standing in for fecal material and allows the intestines to develop to the point where they can process milk immediately on birth. This material is passed by the child in the first few days.

Immediately after birth, a newborn's skin is often grayish to dusky blue in color. As soon as the newborn begins to breathe, usually within a minute or two, the skin's color reaches its normal tone. Newborns are wet, covered in streaks of blood, and coated with a white substance known as vernix caseosa, which is hypothesised to act as anantibacterial barrier. The newborn may also have Mongolian spots, various otherbirthmarks, or peeling skin, particularly on the wrists, hands, ankles, and feet.

A newborn's genitals are enlarged and reddened, with male infants having an unusually large scrotum. The breasts may also be enlarged, even in male infants. This is caused by naturally occurring maternal hormones and is a temporary condition. Females (and even males) may actually discharge milk from their nipples (sometimes called witch's milk), and/or a bloody or milky-like substance from the vagina. In either case, this is considered normal and will disappear in time.

The umbilical cord of a newborn is bluish-white in color. After birth, the umbilical cord is normally cut, leaving a 1–2 inch stub. The umbilical stub will dry out, shrivel, darken, and spontaneously fall off within about 3 weeks. Occasionally, hospitals may apply triple dye to the umbilical stub to prevent infection, which may temporarily color the stub and surrounding skin purple.

Internal physiological changes at birth

Upon entry into the air-breathing world, without the nutrition and oxygenation via the umbilical cord, the newborn must begin to adjust to life outside the uterus. Newborns can feel all different sensations, but respond most enthusiastically to soft stroking, cuddling and caressing. Gentle rocking back and forth often calms a crying infant, as do massages and warm baths. Newborns may comfort themselves by sucking their thumb, or a pacifier. The need to suckle is instinctive and allows newborns to feed.

Newborn infants have unremarkable vision, being able to focus on objects only about 18 inches (45 cm) directly in front of their face. While this may not be much, it is all that is needed for the infant to look at the mother's eyes or areola when breastfeeding. Depth perception does not develop until the infant is mobile. Generally, a newborn cries when wanting to feed. When a newborn is not sleeping, or feeding, or crying, he or she may spend a lot of time staring at various objects. Usually anything that is shiny, has sharp contrasting colors, or has complex patterns will catch an infant's eye. However, the newborn has a preference for looking at other human faces above all else.

In utero, the infant can hear many internal noises, such as the mother's heartbeat, in addition to external noises including the human voice, music and most other sounds. Therefore, though a newborn's ears may have some catarrh and fluid, he or she can hear sound from before birth. Newborns usually respond more readily to a female voice over a male voice. This may explain why people will unknowingly raise the pitch of their voice when talking to newborns (this voice change is called motherese). The sound of other human voices, especially the mother's, can have a calming or soothing effect on the newborn. Conversely, loud or sudden noises will startle and scare them. Newborns have been shown to prefer sounds that were a regular feature of their prenatal environment, for example, the theme tune of a television programme their mother watched regularly.

Newborns can respond to differing tastes, including sweet, sour, bitter, and salty substances, with a preference toward sweetness. It has been shown that neonates show a preference for the smell of foods that their mother ate regularly.

Care and feeding

Infants cry as a form of basic instinctive communication. A crying infant may be trying to express a variety of feelings including hunger, discomfort, overstimulation, boredom, wanting something, or loneliness.

Breastfeeding is the recommended method of feeding by all major infant health organizations. If breastfeeding is not possible or desired, bottle feeding is done with expressed breast-milk or with infant formula. Infants are born with a sucking reflex allowing them to extract the milk from the nipples of the breasts or the nipple of the baby bottle, as well as an instinctive behaviour known as *rooting* with which they seek out the nipple.

Sometimes a wet nurse is hired to feed the infant, although this is rare, especially in developed countries.

Adequate food consumption at an early age is vital for an infant's development. From birth to four months, infants should consume breast milk or an unmodified milk substitute. As an infant's diet matures, finger foods may be introduced as well as fruit, vegetables and small amounts of meat.

As infants grow, food supplements are added. Many parents choose commercial, ready-made baby foods to supplement breast milk or formula for the child, while others adapt their usual meals for the dietary needs of their child. Whole cow's milk can be used at one year, but lower-fat milk should not be provided until the child is 2 to 3 years old. Weaning is the process through which breast milk is eliminated from the infant's diet through the introduction of solid foods in exchange for milk.

Until they are toilet-trained, infants in industrialized countries wear diapers. The transition from diapers to cloth undergarments is an important transition in the development of a infant/baby to that of a toddler. Children need more sleep than adults—up to 18 hours for newborn babies, with a declining rate as the child ages. Until babies learn to walk, they are carried in the arms, held in slings or baby carriers, or transported in baby carriages or strollers. Most industrialized countries have laws requiring child safety seats for babies in motor vehicles.

Infants and the Benefits of Touch

Studies have shown that infants who have been the recipients of positive touch experience more benefits as they develop emotionally and socially. Experiments have done with infants up to four months of age using both positive touch (stroking or cuddling) and negative touch (poking, pinching or tickling). The infants who received the positive touch cried less often and also vocalized and smiled more than the infants who were touched negatively. Infants who were the recipients of negative touching have been linked with emotional and behavioural problems later in life. A lower amount of physical violence in adults has been discovered in cultures with greater levels of positive physical touching.

Infant mortality

Infant mortality is the death of an infant in the first year of life, often expressed as the number of deaths per 1000 live births (infant mortality rate). Major causes of infant mortality include dehydration, infection, congenital malformation and SIDS.

This epidemiological indicator is recognized as a very important measure of the level of health care in a country because it is directly linked with the health status of infants, children, and pregnant women as well as access to medical care, socioeconomic conditions, and public health practices.

It is shown through data that infant mortality rates continue to be very high in the US compared to infant mortality rates in other countries such as Sweden and Japan. There is a positive relationship between national wealth and good health. The rich and industrialized countries of the world, prominently Canada, the United States, and Japan, spend a large proportion of their wealthy budget on the health care system. As, a result, their health care systems are very sophisticated, with many physicians, nurses, and other health care experts servicing the population. Thus, infant mortality is low. On the other hand, a country such as Mexico, which spends disproportionately less of its budget on healthcare, suffers from high mortality rates. This is because the general population is likely to be less healthy. Infant mortality rates are especially high in minority groups. Non-Hispanic black woman have an infant mortality rate of 13.63 per 1000 live births whereas in non-Hispanic white woman it was much lower at a rate of 5.76 per 1000 live births.

Emotional development

Attachment theory is primarily an evolutionary and ethological theory whereby the infant or child *seeks proximity* to a *specified attachment* situations of alarm or distress for the purpose of survival. The forming of attachments is considered to be the foundation of the infant/child's capacity to form and conduct relationships throughout life. Attachment is not the same as love and/ or affection although they often go together. Attachment and attachment behaviours tend to develop between the age of 6 months and 3 years. Infants become attached to adults who are sensitive and responsive in social interactions with the infant, and who remain as consistent caregivers for some time. Parental responses lead to the development of patterns of attachment which in turn lead to 'internal working models' which will guide the individual's feelings, thoughts, and expectations in later relationships. There are a number of attachment 'styles' namely 'secure', 'anxious-ambivalent', 'anxious-avoidant', (all 'organized') and 'disorganized', some of which are more problematic than others. A lack of attachment or a seriously disrupted capacity for attachment could potentially amount to serious disorders.

Baby and plane travel

Many airlines refuse boarding for all babies aged under 7 days (for domestic flights) or 14 days for international flights. Asiana Airlinesallows babies to board international flights at 7 days of age. Garuda Indonesia disallows all babies under the age of 14 days to board any flights. Delta Airlines and Pinnacle Airlines allow infants to travel when they are less than 7 days old when they present a physician travel approval letter. Comair requires a letter if the infant is 12 days or younger. Skywest will not allow an infant less than 8 days old on board.

Adults

In people aged 18 years or older hypertension is defined as a systolic and/ or a diastolic blood pressure measurement consistently higher than an accepted normal value (currently 139 mmHg systolic, 89 mmHg diastolic: see table — Classification (JNC7)). Lower thresholds are used (135 mmHg systolic or 85 mmHg diastolic) if measurements are derived from 24-hour ambulatory or home monitoring. Recent international hypertension guidelines have also created categories below the hypertensive range to indicate a continuum of risk with higher blood pressures in the normal range. JNC7 (2003) uses the term prehypertension for blood pressure in the range 120-139 mmHg systolic and/or 80-89 mmHg diastolic, while ESH-ESC Guidelines (2007) and BHS IV (2004) use optimal, normal and high normal categories to subdivide pressures below 140 mmHg systolic and 90 mmHg diastolic.

Hypertension is also sub-classified: JNC7 distinguishes hypertension stage I, hypertension stage II, and isolated systolic hypertension. Isolated systolic hypertension refers to elevated systolic pressure with normal diastolic pressure and is common in the elderly. The ESH-ESC Guidelines (2007) and BHS IV (2004), additionally define a third stage (stage III hypertension) for people with systolic blood pressure exceeding 179 mmHg or a diastolic pressure over 109 mmHg. Hypertension is classified as "resistant" if medications do not reduce blood pressure to normal levels.

Neonates and Infants

Hypertension in neonates is rare, occurring in around 0.2 to 3% of neonates, and blood pressure is not measured routinely in the healthy newborn. Hypertension is more common in high risk newborns. A variety of factors, such as gestational age, postconceptional age and birth weight needs to be taken into account when deciding if a blood pressure is normal in a neonate.

Children and Adolescents

Hypertension occurs quite commonly in children and adolescents (2-9% depending on age, sex and ethnicity) and is associated with long term risks of ill-health. It is now recommended that children over the age of 3 have their blood pressure checked whenever they attend for routine medical care or checks, but high blood pressure must be confirmed on repeated visits before characterizing a child as having hypertension. Blood pressure rises with age in childhood and, in children, hypertension is defined as an average systolic or diastolic blood pressure on three or more occasions equal or higher than the 95th percentile appropriate for the sex, age and height of the child. Prehypertension in children is defined as average systolic or diastolic blood pressure that is greater than or equal to the 90th percentile, but less than the 95th percentile. In adolescents, it has been proposed that hypertension and pre-hypertension are diagnosed and classified using the same criteria as in adults.

Signs and Symptoms

Hypertension is rarely accompanied by any symptoms, and its identification is usually through screening, or when seeking healthcare for an unrelated problem. A proportion of people with high blood pressure reports headaches (particularly at the back of the head and in the morning), as well as lightheadedness, vertigo, tinnitus (buzzing or hissing in the ears), altered vision or fainting episodes.

On physical examination, hypertension may be suspected on the basis of the presence of hypertensive retinopathy detected by examination of the optic fundus found in the back of the eye using ophthalmoscopy. Classically, the severity of the hypertensive retinopathy changes is graded from grade I–IV, although the milder types may be difficult to distinguish from each other. Ophthalmoscopy findings may also indicate how long a person has been hypertensive.

ADOLESCENT

The World Bank and its partners have been challenged anew to work with countries to tackle un-addressed health and development problems of vulnerable groups. Among these groups, adolescents have been identified as particularly important, and often neglected. There is increasing concern about the negative consequences of malnutrition in adolescent girls and young women for their offspring and themselves. Renewed action to address the nutritional problems and needs of adolescents is therefore a priority.

There are 1.2 billion adolescents between the ages of 10 and 19 in developing nations, making up one fifth to one quarter of their populations. While adolescents have typically been considered a low risk group for poor health, this ignores the fact that many health problems later in life can be improved or avoided by adopting healthy lifestyle habits in adolescence. There is substantial evidence that inadequate diets affect adolescents' ability to learn and work at maximum productivity. Undernutrition increases the risk of poor obstetric outcomes for teen mothers and jeopardizes the healthy development of their future children. Children born to short, thin women are more likely themselves to be stunted, underweight and less cognitively able than normal birthweight peers. In addition, the heightened obstetric risk caused by stunting in childhood and adolescence persists throughout a woman's reproductive life.

Adolescence is also a unique intervention point in the life-cycle. It is a stage of receptivity to new ideas and a point at which lifestyle choices may determine an individual's life course. It offers a chance to acquire knowledge about optimal nutrition during young adulthood that could prevent or delay adult-onset diet-related illnesses later on. Potentially, the inclusion of adolescent boys in nutrition and healthy lifestyle programs will contribute to the improved nutrition and health of women during childbearing and for infants and young children in the critical early years of life.

Attention needs to be directed at the link between adolescent nutrition and immediate and long-term health issues, including the cost effectiveness of addressing adolescent nutrition, so that the political commitment to support an action agenda can be secured.

Nutrition for adult men

Adulthood is the period of strength, virility, activity and responsibility for both men and women. Men have the responsibility of earning the lion's share of income to provide for their families, protect them and provide life to future generations. A healthy man is able to discharge his duties well, and optimal nutrition is essential to maintain good health. In adulthood, nutrition is required to perform various functions of the body such as to provide the body with energy for day-to-day activities, promote replacement and regeneration of worn-out tissues, maintain basic metabolism and so on. Adequate amount and adequate type of nutrition promotes a healthy weight, prevents diseases and contributes to the overall well-being of the individual.

Our contemporary lifestyle hardly gives us the time and energy to adhere to a healthy eating and exercise regime. In this fast paced world, work, competition, family responsibilities and pressure often take precedence and leave us space and time only for erratic eating behaviours and an inactive lifestyle. So, what's the solution to the problem of the consort of imbalanced nutrition and lack of physical activity?

Increase healthy eating options

The answer is simple: gather information on all possible healthy eating options available (and viable for you) and create your own daily diet plan. If you are overweight, decrease the portions or serving sizes of the food items and increase your daily activity level.

Increase your activity level

Perform simple exercises whenever you have the time. For example, whenever there's a need to fetch something like water, make it a habit to get it yourself. You can walk while talking on your mobile or listening to music. Take the stairs instead of the lift. In short, workout whenever feasible.

Understand your nutritional requirements

In order for you to understand your body's nutritional needs, you can rely on the food pyramid guide we have provided in this article. Information in this article will also make you understand the concept of the different food groups and the amount in which they have to be taken.

Nutritional needs of adult men

For healthy adult men (without any medical condition and aged between 20 and 39 years), energy requirements are based on height, weight and activity levels. Imagine that your daily energy requirements are 2200 calories, but you take in around 2500 calories, that's about 300 calories extra per day; you will store this in the form of fat in your body. On the other hand, if you take in 2000 calories per day, that's 200 calories less than that required by your body, the extra energy requirements are met by burning glycogen and later fat stored inside your body. The idea here is to balance the calorie intake with the amount of calories burnt in order to maintain a healthy weight. However, to shed weight, you will have to take in fewer calories than needed by your body consistently. Similarly, to gain weight, you will have to consistently take in more calories than you burn.

Calculating your daily calorie requirements

Energy requirements are the sum of the energy needed to carry out basic metabolic functions and energy requirements for daily physical activity. Both of these depend upon the weight of the person.

For example, let us calculate the energy requirements of a man weighing 60 kg. For each kg of weight, 1.3 calories per hour on an average are required to meet the basic metabolic rate (BMR). Therefore, the daily calorie requirements to meet the basic metabolic needs will be

Calorie requirements for meeting BMR=1.3 (calories) × 24 (hours) × 60 (kg) =1872 calories/day.

Now, let us calculate the energy requirements for performing daily physical activities. This depends upon the duration and intensity of activities carried out daily. For example, for performing medium-intensity exercise, a person normally burns 8.5 calories per hour per kg body weight. So, if you perform such exercises an hour a day, you burn

Calorie requirements for physical activity= 8.5 (calories) × 1 (hour) × 60 (kg) = 510 calories.

Therefore,

Your daily energy requirements = Calorie requirements for meeting BMR + Calorie requirements for physical activity = 1872 + 510 = 2382 calories.

The above food pyramid is useful to get an idea of what components are essential for proper nutrition and in roughly what amounts they have to be taken in a day. As you can see in the diagram, the base of the pyramid (and the largest section) is occupied by the food group 'Grains and cereals', signifying that these should form the main source of energy for an adult man. As we go up, we find two more sections, the "Fruit group" and the "Vegetable Group". These two food groups are also essential for adequate nutrition and must form a component of the daily diet plan.

As we progress up further, we come to the groups "Milk and milk products group" and the "Meat, nuts and beans group". The former food group is vital to meet the calcium, phosphorus and mineral requirements of the body while the latter supplies us with proteins. The apex of the pyramid has "Fats, oils and sugar group" as the last food group. The smallest size of this pyramid shows that these food items should not be the major source of energy and their intake should be limited.

To understand how to improve your diet further, you can refer to the table below which gives examples of the components in the different food groups, the recommended serving size and the number of servings required to be taken per day.

Guidelines for healthy eating

Busy executives can have a tough time keeping their meal routines in sync with their daily requirements of nutrients. They frequently find themselves bingeing on fast food, unhealthy high fat snacks and aerated beverages. One of the chief reasons for this is not planning your daily diet or eating out of stress. The first step to follow is to make sure you don't skip your meals and take them on time. If you find yourself hungry in between, grab a quick bite of whole fruit and drink a milk beverage without added sugar. Remember to start your day with a good wholesome breakfast. Skipping breakfast makes the metabolism sluggish; this causes you to end up eating more during the rest of the day.

The second step is to eat smart. Opt for high-fibre foods such as whole grains and unprocessed cereals such as sprouts and rotis made of whole-wheat flour instead of pasta made of refined flour. This provides bulk to your diet, without the calories and also promotes the health of your digestive system. Refrain eating meals outside or opting for take-away foods, as most of these options tend to be calorie-dense, packed with loads of fats and they don't seem to light on your wallet too! Whenever you eat out, opt for the healthier varieties: grilled or steamed foods, salads minus their dressings, plain whole wheat *rotis* instead of butter *naan, parathas and kulchas*; stir-fried vegetable-based recipes instead of heavy gravies cooked in dry fruits; *fruit chaat* instead of ice-creams.

The third step is to restrict your sugar intake. Avoid aerated beverages that are high on sugar. Make it a habit to go for smarter options such as buttermilk or coconut water to quench your thirst. Use honey instead of sugar in your desserts, if possible.

The fourth is to reduce your calorie-intake from low-nutrient foods. Consume less coffee, tea, potato chips and fried stuff.

The fifth step is to drink plenty of water, around 10 to 12 glasses during the day, even while at work

AGING-GERIATRIC

The Aging Body

Physiological changes occur slowly over time in all body systems. These changes are influenced by life events, illnesses, genetic traits and socioeconomic factors.

Sensory Changes

Sensory changes include a decline in sight and peripheral vision, hearing, smell and taste. The losses are neither total nor rapid, but they do affect nutritional intake and health status.

- Loss of visual acuteness may lead to less activity or a fear of cooking, especially using a stove. Inability to read food prices, nutrition labels or recipes may affect grocery shopping, food preparation and eating. This could have an adverse effect on nutritional status.
- Loss of hearing may lead to less eating out or not asking questions of the waiter or store clerk.
- Changes in smell and taste are more obvious. If food doesn't taste appetizing or smell appealing, we don't want to eat it. If we must cut back on salt, sugar or fat, we may tend not to eat.

Structural Changes

As we age, we lose lean body mass. Reduced muscle mass includes skeletal muscle, smooth muscle and muscle that affects vital organ function, with loss of cardiac muscle perhaps the most important. Cardiac capacity can be reduced and cardiac function impaired by chronic diseases such as atherosclerosis, hypertension or diabetes. Changes also occur in the kidneys, lungs and liver, and in our ability to generate new protein tissue. In addition, aging can slow the immune system's response in making antibodies. The most significant result of the loss of lean body mass may be the decrease in basal energy metabolism. Metabolic rate declines proportionately with the decline in total protein tissue. To avoid gaining weight, we must reduce calorie intake or increase activity. The goal is energy balance.

Loss of lean body mass also means reduced body water — 72 percent of total body water is in lean muscle tissue.

Total body fat typically increases with age. This often can be explained by too many calories. As we age, fat tends to concentrate in the trunk and as fat deposits around the vital organs. However, in more advanced years, weight often declines.

Finally, we lose bone density. After menopause, women tend to lose bone mass at an accelerated rate. Recent attention has focused on the high incidence of osteoporosis. Severe osteoporosis is debilitating and serious. Fractures and their associated illness and mortality are certainly a concern. Also, vertebral compression fractures can change chest configuration. This, in turn, can affect breathing, intestinal distension and internal organ displacement.

Nutrition can be a factor in all of the changes noted above. However, the slowing of the normal action of the digestive tract plus general changes have the most direct effect on nutrition. Digestive secretions diminish markedly, although enzymes remain adequate. Adequate dietary fibre, as opposed to increased use of laxatives, will maintain regular bowel function and not interfere with the digestion and absorption of nutrients, as occurs with laxative use or abuse.

SUGGESTIONS FOR COPING WITH CHANGE

Sensory changes

Loss of smell and taste affect the nutritional intake and status of many seniors. If food does not smell or taste appetizing, it will not be eaten.

Suggestions:

- Try a variety of new food flavors. Experiment with low sodium seasonings such as lemon juice, dill, curry and herbs of all types.
- Sometimes the problem is not dulled senses, but rather a drab, soft diet. Don't cook vegetables until they are mushy. Instead, reawaken the senses to fresh, flavorful foods and new textures

Loss of teeth

Improperly fitting dentures may unconsciously change eating patterns because of difficulty with chewing. A soft, low-fibre diet without important fresh fruits and vegetables may result.

Suggestions:

- Have poorly fitting dentures adjusted.
- Chop, steam, stew, grind or grate hard or tough foods to make them easier to chew without sacrificing their nutritional value. Try a grated carrot and raisin salad.

Osteoporosis

Weight-bearing exercise and a diet high in calcium help protect against osteoporosis. Current treatments include estrogen replacement, exercise and calcium supplements.

Suggestions:

- Walk, lift weights, swim or enroll in a group fitness or water aerobics class. Exercise at least three times a week and have fun!

- Include two to four daily servings of dairy products such as milk, yogurt or cheese.
- If digesting milk is a problem, cultured dairy products, like buttermilk and yogurt, often are tolerated well. Use lactaid, available in most stores, to make reduced lactose milk.
- Post-menopausal women may need a calcium supplement if they can't get enough through diet alone. Talk to a physician or registered dietitian.

Specific Nutrient Needs

Calorie needs change due to more body fat and less lean muscle. Less activity can further decrease calorie needs. The challenge for the elderly is to meet the same nutrient needs as when they were younger, yet consume fewer calories.

The answer to this problem is to choose foods high in nutrients in relation to their calories. Such foods are considered "nutrient-dense." For example, low-fat milk is more nutrient dense than regular milk. Its nutrient content is the same, but it has fewer calories because it has less fat.

Protein needs usually do not change for the elderly, although research studies are not definitive. Protein requirements can vary because of chronic disease. Balancing needs and restrictions is a challenge, particularly in health care facilities. Protein absorption may decrease as we age, and our bodies may make less protein. However, this does not mean protein intake should be routinely increased, because of the general decline in kidney function. Excess protein could unnecessarily stress kidneys.

Reducing the overall fat content in the diet is reasonable. It is the easiest way to cut calories. This is appropriate to reduce weight. Lower fat intake is often necessary because of chronic disease.

About 60 percent of calories should come from carbohydrates, with emphasis on complex carbohydrates. Glucose tolerance may decrease with advancing years. Complex carbohydrates put less stress on the circulating blood glucose than do refined carbohydrates.

Such a regime also enhances dietary fibre intake. Adequate fibre, together with adequate fluid, helps maintain normal bowel function. Fibre also is thought to decrease risk of intestinal inflammation. Vegetables, fruits, grain products, cereals, seeds, legumes and nuts are all sources of dietary fibre.

Vitamins and Minerals

Vitamin deficiencies may not be obvious in many older people. However, any illness stresses the body and may be enough to use up whatever stores there are and make the person vitamin deficient. Medications also interfere with many vitamins. When drug histories are looked at, nutrient deficiencies emerge. Eating nutrient-dense foods becomes increasingly important when calorie needs decline but vitamin and mineral needs remain high. The body can store fat-soluble vitamins and usually the elderly are at lower risk of fat-soluble vitamin deficiencies. Always provide vitamin D-fortified milk for the housebound, nursing home residents, and anyone who does not get adequate exposure to sunlight. Iron and calcium intake sometimes appears to be low in many elderly. To improve absorption of iron, include vitamin C-rich fruits and vegetables with these foods. For example, have juice or sliced fruit with cereal, a baked potato with roast beef, vegetables with fish, or fruit with chicken. To boost your intake of calcium, have tomato slices in a cheese sandwich, or salsa with a bean burrito.

Zinc can be related to specific diseases in the elderly. It can also be a factor with vitamin K in wound healing. Zinc improves taste acuity in people where stores are low. Habitual use of more than 15 mg per day of zinc supplements, in addition to dietary intake, is not recommended without medical supervision. If you eat meats, eggs and seafood, zinc intake should be adequate. This underscores the importance of eating a wide variety of foods.

Zinc along with vitamins C and E, and the phytochemicals lutein, zeaxanthin and beta-carotene may help prevent or slow the onset of age-related macular degeneration. The best way to obtain these nutrients is to consume at least five servings of fruits and vegetables, especially dark green, orange and yellow ones. Good choices include kale, spinach, broccoli, peas, oranges and cantaloupes.

Vitamin E may have a potential role in the prevention of Alzheimer's disease. Research has shown that eating foods with vitamin E, like whole grains, peanuts, nuts, vegetable oils, and seeds, may help reduce the risk of Alzheimer's disease. However, the same benefits did not hold true for vitamin E from supplements. Low levels of vitamin B12 have been associated with memory loss and linked to age-related hearing loss in older adults. Folate, which is related to B12 metabolism in the body, may actually improve hearing. However, if B12 levels are not adequate, high folate levels may be a health concern. As we age, the amount of the chemical in the body, needed to absorb vitamin B12 decreases. To avoid deficiency, older adults are advised to eat foods rich in vitamin B12 regularly, including meat, poultry, fish, eggs and dairy foods. Consult your doctor to see if a vitamin B12 supplement may also be necessary.

Drugs used to control diseases such as hypertension or heart disease can alter the need for electrolytes, sodium and potassium. Even though absorption and utilization of some vitamins and minerals becomes less effective with age, higher intakes do not appear to be necessary. As for any age group, it's important to enjoy a wide variety of foods. Also, if you are taking an herbal or dietary supplement, make sure to tell your doctor, since these supplements may interact with other drugs or nutrients in your diet.

Water

Generally, water as a nutrient receives little attention once a person is old enough to talk. However, of all the nutrients, water isthe most important, serving many essential functions.

Adequate water intake reduces stress on kidney function, which tends to decline with age. Adequate fluid intake also eases constipation. With the aging process, the ability to detect thirst declines, so do not wait to drink water until you are thirsty. Drink plenty of water, juice, milk, and coffee or tea to stay hydrated. Drink the equivalent to five to eight glasses everyday. It may be helpful to use a cup or water bottle which has calibrated measurements on it, in order to keep track of how much you drink. Carry it with you throughout your home or wherever you go during the day.

Variety of Foods

People of all ages need more than 40 nutrients to stay healthy. With age, it becomes more important that diets contain enough calcium, fibre, iron, protein, and the vitamins A, C, D and Folacin. Reduce calories, select nutrient-dense foods, and enjoy smaller portions of foods high in fat, sugar and sodium. Because no one food or pill provides all of the nutrients, eat a variety of foods to get the full spectrum of nutrients.

Variety often is lacking in the diets of seniors, who often eat the same foods over and over again. Get out of this food rut by trying some of these suggestions:

- Eat breakfast foods for lunch or lunch foods for dinner.
- Use color as a guide for variety in a meal. A good meal should provide three distinct colors on the plate.
- Increase the variety of texture in meals. Add whole grain breads (rye, wheat, pumpernickel), whole grain cereals, and cooked legumes (beans of all types, lentils, dried peas).
- Eat at least five servings of fruits and vegetables each day.
- Build your diet as MyPyramid recommends.

Eating Alone

Pamper yourself and pay special attention to meal preparation. If eating alone means skipping meals or not eating a variety of foods, try one of the following to add mealtime sparkle:

- Start an eating club.
- Eat by a window and use the best dishes.
- Eat a lunch in the park.
- Treat yourself to a meal out.
- Invite a friend to a potluck dinner.
- Attend the nutrition program for the elderly and enjoy meals in the community.
- During an illness, arrange for home-delivered meals from the community nutrition program for the elderly.
- Make leftovers into "planned leftovers."
- Prepare a new recipe each week and invite friends over for a tasting party.
- Use frozen prepared dinners for added variety and convenience.
- Add a special touch to your table, like a colorful table cloth or placemats, a plant, candle, or decorative vase.

Buy More Nutrients Per Food Dollar

On any budget, it requires planning to buy a variety of nutrient-rich foods. You get more nutrients for your dollar from milk, eggs, legumes, grain, lean meat, fish or poultry than from prepared processed foods, ready-made desserts and snack foods.

- Look for specials. Check the newspaper ads for sales on meat, fish, poultry, fruits and vegetables.
- Use coupons. Coupons save money on products you usually buy. Buying expensive convenience foods with coupons may cost more.
- Plan ahead. Buy an advertised meat special to use for several meals. Freeze portions for later use. Keep a supply of food on hand for bad weather or illness. Also, buy a few gallons of water to keep on hand in case of an emergency.
- Make a shopping list. Plan meals for the week, then make a list of the food needed.
- Shopping. Shop when the store is not crowded and after a meal. Ask for help at the store, if needed, or shop with a friend.
- Shop on the telephone. Many stores will deliver your order if you are unable to go to the store. Check what is available in your community.
- Read the labels. Find out the ingredients in the food and its expiration date. If you can't find the date, ask for the information. If enough people ask, it will appear in larger print.
- Read the nutrition labels. When the label says "low calorie" or "low sodium," check the nutrition information label for the number of

calories and amount of sodium per serving. Sodium labels give valuable information on the sodium content per serving: sodium free, 5 mg; very low sodium, 35 mg; low sodium, 140 mg; reduced sodium, 75 percent less sodium than before. Unsalted foods or foods processed without salt may still contain sodium. If necessary take a magnifying glass to read the fine print on the labels.

- Use unit pricing. Unit pricing helps you find the best buy. The unit pricing label tells the cost per unit of measure such as ounce, serving or pound. Use this to compare brands and different size packages to get the most food per dollar.

Interactions of the Origins

The nutritional need originates in the chemical reactions that obey the principles of thermodynamics, and arose during the billion years following the formation of the earth. The different energy sources that existed at the time enabled chemical compounds such as water and water vapour (H_2O), methane (CH_4) and ammonia (NH_3) to interact and combine, thus generating the basic components of living matter. Laboratory simulations of the chemical and energy conditions assumed to have prevailed at the beginning of the earth have permitted the recreation of practically all these basic components from these three simple gases. The interaction then continued through the condensation of the basic components in long chains that generated the molecules (polymers) typical of the living world, and later the assembly of polymers in organelles – these in turn formed living cells. It took a billion years for the first single-celled forms of life (or unicellulars) that compose the bacterial kingdom to come into being. It took a further three billion years for unicellulars to develop, and later assemble in, multi-cellular organisms, with the differentiation of organs, and so give rise to the vegetable and animal kingdoms. Finally, the three kingdoms continued to develop during the next 800 million years or so, by fashioning the earth and its atmosphere to bring them approximately to their contemporary form. *Homo sapiens* (or, rather, *Homo economicus*, an expression that better reflects his exploitation of the environment) appeared only one hundred thousand years ago. During this evolution, the principles of thermodynamics remained a denominator common to the original energy reactions and the nutritional need. The universal principles that govern the exchanges of energy.

VEGAN NUTRITION

A vegan is a strict vegetarian who does not eat any dairy products or eggs. Most vegans do not eat honey. A well balanced vegan diet can provide all the essential nutrients you require and shares the same health advantages as a vegetarian diet. Nutritional guidelines for vegans are essentially similar to those for vegetarians. However, vegetarians gain certain nutrients from dairy products and eggs. Vegans need to ensure their diets contain plant food sources of these nutrients.

Protein

Obtaining adequate protein on a vegan diet is not a problem. Nuts & seeds, pulses, wholegrain and grain products and soya products all supply protein. Previously, it has been thought that plant proteins are of a lower quality than animal proteins in terms of their essential amino acid content. However, this is no longer regarded as a problem and eating a balanced diet of plant foods will provide all the essential amino acids in adequate amounts.

Essential Fatty Acids

There are two essential fatty acids which must be supplied by the diet. These are linoleic acid and a-linolenic acid. Essential fatty acids are important for cell membrane function, cholesterol metabolism and the synthesis of various metabolites. Good sources of essential fatty acids are vegetable oils. It is important to have the correct balance between linoleic acid and a-linolenic acid. It has been suggested that vegans should use soyabean or rapeseed oils rather than sunflower or corn oils as these help give a better dietary balance.

Vitamin B2 (Riboflavin): Certain studies have found vegans to have a low intake of the vitamin, riboflavin. Riboflavin is important in converting protein, fats and carbohydrates into energy, and the synthesis and repair of body tissues. Good sources of riboflavin include whole grains, mushrooms, almonds, leafy green vegetables and yeast extracts.

Vitamin B12: Vitamin B12 is found primarily in meat, dairy products and eggs and is absent from plant foods. Considerable research has been carried out into possible plant sources of B12. Fermented soya products, seaweeds and algae such as spirulina have all been proposed as containing significant amounts of B12. However, the present consensus is that any B12 present in plant foods is likely to be in a form unavailable to humans and so these foods should not be relied upon as safe sources.

Vitamin B12 is important in the formation of red blood cells and the maintenance of a healthy nervous system. When deficiency does occur it is more likely to be due to a failure to absorb B12 from the intestine than a dietary deficiency.

Vegans can obtain B12 from a wide range of foods which have been fortified with the vitamin. These include certain yeast extracts, veggieburger mixes, breakfast cereals, vegetable margarines and soya milks. You should check the packaging to see which individual products are fortified with B12.

Vitamin D: Vitamin D is present in oily fish, eggs and dairy products in variable amounts. It is not found in plant foods. However, vegans can obtain vitamin D from vegetable margarines, some soya milks and certain other foods which are fortified with the vitamin.

Vitamin D is also synthesised by the skin when exposed to sunlight. Synthesis of vitamin D in this way is usually adequate to supply all the body's requirements. Most vegans will obtain sufficient vitamin D providing they spend time outdoors on bright days. Fortified foods further ensure adequate amounts.

Vegans who may be confined indoors may be recommended a vitamin D supplement. Also, infants who are seldom outdoors or who are dark-skinned may require supplements. Asian vegans may also be at risk of deficiency, particularly Asian women who may be required to keep their skin covered for cultural reasons.

Calcium: The major source of calcium in British diets is generally milk and dairy products. Vegans can obtain adequate calcium from plant foods. Good sources include tofu, leafy green vegetables, watercress, dried fruit, seeds and nuts. Also, white bread is fortified with calcium, as are some soya milks. Hard water can also provide significant amounts of calcium.

Iodine: Milk is the primary source of iodine in the British diet and studies have indicated some vegans may have a low iodine intake. Seaweeds are a good source of iodine, and vegetables and grains can contain iodine depending on the amounts in the soil.

Infants: It is perfectly possible to bring up a child on a vegan diet. Vegan children should be given plenty of nutrient rich foods and need good sources of protein, calcium, vitamin B12 and vitamin D. High fibre foods can fill up a child without filling their nutritional needs as well as interfering with mineral absorption from the intestine. For these reasons, foods high in fibre shouldn't be overused.

Vegan Storecupboard: Dairy products can largely be replaced with various soya products. There are several brands of soya milk. It can be purchased either sweetened or unsweetened, plain or flavoured. Different brands may be fortified with vitamin B12, vitamin D and calcium. Soya cheeses, yoghurts and cream are all available from health food stores. Eggs can be replaced in recipes by commercial egg replacer products, also available from health food stores.

Vitamin B12

Vitamin B12 is a member of the vitamin B complex. It contains cobalt, and so is also known as cobalamin. It is exclusively synthesised by bacteria and is found primarily in meat, eggs and dairy products.

There has been considerable research into proposed plant sources of vitamin B12. Fermented soya products, seaweeds, and algae such as spirulina have all been suggested as containing significant B12. However, the present consensus is that any B12 present in plant foods is likely to be unavailable to humans and so these foods should not be relied upon as safe sources. Many vegan foods are supplemented with B12.

Vitamin B12 is necessary for the synthesis of red blood cells, the maintenance of the nervous system, and growth and development in children. Deficiency can cause anaemia. Vitamin B12 neuropathy, involving the degeneration of nerve fibres and irreversible neurological damage, can also occur.

Functions

Vitamin B12's primary functions are in the formation of red blood cells and the maintenance of a healthy nervous system. B12 is necessary for the rapid synthesis of DNA during cell division. This is especially important in tissues where cells are dividing rapidly, particularly the bone marrow tissues responsible for red blood cell formation. If B12 deficiency occurs, DNA production is disrupted and abnormal cells called megaloblasts occur. This results in anaemia. Symptoms include excessive tiredness, breathlessness, listlessness, pallor, and poor resistance to infection. Other symptoms can include a smooth, sore tongue and menstrual disorders. Anaemia may also be due to folic acid deficiency, folic acid also being necessary for DNA synthesis.

B12 is also important in maintaining the nervous system. Nerves are surrounded by an insulating fatty sheath comprised of a complex protein called myelin. B12 plays a vital role in the metabolism of fatty acids essential for the maintenance of myelin. Prolonged B12 deficiency can lead to nerve degeneration and irreversible neurological damage. When deficiency occurs, it is more commonly linked to a failure to effectively absorb B12 from the intestine rather than a dietary deficiency. Absorption of B12 requires the secretion from the cells lining the stomach of a glycoprotein, known as intrinsic factor. The B12-intrinsic factor complex is then absorbed in the ileum (part of the small intestine) in the presence of calcium. Certain people are unable to produce intrinsic factor and the subsequent pernicious anaemia is treated with injections of B12.

Vitamin B12 can be stored in small amounts by the body. Total body store is 2-5mg in adults. Around 80% of this is stored in the liver. Vitamin B12 is excreted in the bile and is effectively reabsorbed. This is known as enterohepatic circulation. The amount of B12 excreted in the bile can vary from 1 to 10ug (micrograms) a day. People on diets low in B12, including vegans and some vegetarians, may be obtaining more B12 from reabsorption than from dietary sources. Reabsorption is the reason it can take over 20 years for deficiency disease to develop in people changing to diets absent in B12. In comparison, if B12 deficiency is due to a failure in absorption it can take only 3 years for deficiency disease to occur.

Dietary Sources

The only reliable unfortified sources of vitamin B12 are meat, dairy products and eggs. There has been considerable research into possible plant food sources of B12. Fermented soya products, seaweeds and algae have all been proposed as possible sources of B12. However, analysis of fermented soya products, including tempeh, miso, shoyu and tamari, found no significant B12. Spirulina, an algae available as a dietary supplement in tablet form, and nori, a seaweed, have both appeared to contain significant amounts of B12 after analysis. However, it is thought that this is due to the presence of compounds structurally similar to B12, known as B12 analogues. These cannot be utilised to satisfy dietary needs. Assay methods used to detect B12 are unable to differentiate between B12 and it's analogues, Analysis of possible B12 sources may give false positive results due to the presence of these analogues.

Researchers have suggested that supposed B12 supplements such as spirulina may in fact increase the risk of B12 deficiency disease, as the B12 analogues can compete with B12 and inhibit metabolism. The current nutritional consensus is that no plant foods can be relied on as a safe source of vitamin B12. Bacteria present in the large intestine are able to synthesise B12. In the past, it has been thought that the B12 produced by these colonic bacteria could be absorbed and utilised by humans. However, the bacteria produce B12 too far down the intestine for absorption to occur, B12 not being absorbed through the colon lining.

Human faeces can contain significant B12. A study has shown that a group of Iranian vegans obtained adequate B12 from unwashed vegetables which had been fertilised with human manure. Faecal contamination of vegetables and other plant foods can make a significant contribution to dietary needs, particularly in areas where hygiene standards may be low. This may be responsible for the lack of aneamia due to B12 deficiency in vegan communities in developing countries.

Good sources of vitamin B12 for vegetarians are dairy products or free-range eggs. ½ pint of milk (full fat or semi skimmed) contains 1.2 μg. A slice of vegetarian cheddar cheese (40g) contains 0.5 μg. A boiled egg contains 0.7 μg. Fermentation in the manufacture of yoghurt destroys much of the B12 present. Boiling milk can also destroy much of the B12.

Vegans are recommended to ensure their diet includes foods fortified with vitamin B12. A range of B12 fortified foods are available. These include yeast extracts, Vecon vegetable stock, veggieburger mixes, textured vegetable protein, soya milks, vegetable and sunflower margarines, and breakfast cereals.

Required Intakes

The old Recommended Daily Amounts (RDA's) have now been replaced by the term Reference Nutrient intake (RNI). The RNI is the amount of nutrient which is enough for at least 97% of the population.

Reference Nutrient Intakes for Vitamin B12, μg/day. (1000 μg = 1mg)

Age	RNI
0 to 6 months	0.3 μg
7 to 12 months	0.4 μg
1 to 3 yrs	0.5 μg
4 to 6 yrs	0.8 μg
7 to 10 yrs	1.0 μg
11 to 14 yrs	1.2 μg
15 + yrs	1.5 μg
Breast feeding women	2.0 μg

Pregnant women are not thought to require any extra B12, though little is known about this. Lactating women need extra B12 to ensure an adequate supply in breast milk.

B12 has very low toxicity and high intakes are not thought to be dangerous.

DEPENDENCY ON LIVING SPECIES

The third level of nutritional dependency developed in parallel with the evolution of species. Specialization, differentiation and adaptation to the environment entailed nutritional specificities that implied unilateral dependencies and, more importantly, symbiotic interdependencies with major consequences for the survival of the living world in general. The interdependency of species can take many different forms that fall into two main categories: dependency on a source of energy or fuel, and dependency on structural or functional matter.

Dependency on matter

As discussed above, energy flows in an organism cause fluxes of matter. The latter flows from one type of organism to another, not necessarily through trophic dependency (downstream-upstream), but according to a cyclical rhythm – in other words, an interdependency. In these cycles, organisms exchange the nutrients required for their survival. Without entering into detail as to the sometimes highly complex nutritional interdependencies that govern the biosphere, this discussion will focus on the three cycles involving practically the entire biosphere: carbon, oxygen and nitrogen. These three elements are found in the atmosphere in the form of simple gases; they can be dissolved in water or combined with solids. Their movement throughout the biosphere is thus not accidental. Furthermore, their physical and chemical features lead to the production of highly diverse molecules, on the one hand, and huge combinations of such molecules, on the other.

The carbon and oxygen cycle

Organisms capable of photosynthesis combine carbonic acid gas with water to generate organic compounds, giving off oxygen in the process. On the other hand, organisms that are incapable of photosynthesis consume organic compounds and oxygen, and give off carbonic acid gas and water in the process of breathing and fermentation. This introduces a carbon and oxygen cycle into the biosphere.

Dependency on energy flow

The first stage of energy flow in the biological world belongs to the realm of photosynthesis that uses solar radiation to create organic compounds, and the latter are the primary source of energy of organisms that are incapable of photosynthesis. The realm of photosynthesis is thus at the basis of all the energy used by the rest of the biosphere. In a way, there exists a form of "energy parasitism" in stages: the world of photosynthesis is a parasite (or predator) of sunlight, and the world of fermentation and breathing is a parasite of photosynthesis. This energy dependency can be illustrated by the chain linking carnivores to herbivores, herbivores to green plants, and green plants to sunlight. In this specific case, later links (carnivores) depend on earlier ones (ultimately, sunlight). Species are thus connected by the nature of their nutritional energy need in a sequence of levels or stages (referred to as trophic levels). Plants represent the first trophic level, herbivores the second, and so on. Each level is a consumer of the previous, and a producer for the following level.

NUTRITION DURING PREGNANCY

The conception and the subsequent weeks afterwards is the time when it is at its most vulnerable, as it is the time when the organs and systems develop within. The energy used to create these systems comes from the energy and nutrients in the mother's circulation, and around the lining of the womb, such is the reason why correct nutrient intake during pregnancy is so important.

During the early stages of pregnancy, since the placenta is not yet formed, there is no mechanism to protect the embryo from the deficiencies which may be inherent in the mother's circulation. Thus, it is critical that an adequate amount of nutrients and energy is consumed. Additionally, the frequent consumption of nutritious foods helps to prevent nausea, vomiting, and cramps. Supplementing one's diet with foods rich in folic acid, such as oranges and dark green leafy vegetables, helps to prevent neural tube birth defects in the baby. Consuming foods rich in iron, such as lean red meat and beans help to prevent anemia and ensure adequate oxygen for the baby. A necessary step for proper diet is to take a daily prenatal vitamins, that ensure their body gets the vitamins and minerals it needs to create a healthy baby. These vitamins contain folic acid, iodine, iron, vitamin A, vitamin D, zinc and calcium.

By pregnancy trimester

First trimester

The first necessary step for proper diet is to take a daily prenatal vitamins, that ensure their body gets the vitamins and minerals it needs to create a healthy baby. These vitamins contain folic acid, iodine, iron, vitamin A, vitamin D, zinc and calcium. Hence the diet should consist of fruits, vegetables, non fat dairy foods, lean meat and whole grains which are rich in the prenatal vitamins and minerals. Eating well is very important during pregnancy, the food they eat keeps them and their baby nurtured. It should be right balance of carbohydrates, fat and protein without too much of calories. Drink fruit juices with 100% fruit and limit them to one glass per day. Juices are not only high in calories but do not provide the amount of fibre compared to a whole fruit. Include at least one serving of citrus fruit like orange, grape fruit as they are rich in vitamin C. Vegetable servings should include a variety of green leafy (broccoli, spinach, kale); red (tomatoes, red peppers); yellow (yellow peppers, corn) and orange vegetables (carrot, sweet potatoes, pumpkin). They should have at least 3 - 4 servings each day of fruits and 3 - 5 servings of vegetables each day during their first trimester. The foetus requires more of vitamins and minerals during the first trimester than extra calories. If the mothers are very hungry there is no harm in eating more food compared to their intake before pregnancy. It is a mandatory for the expecting mothers who by their obstetrician are very thin. Organic fat and protein build healthy babies. Succumbing to junk food will only deplete them of vital nutrients which is not at all advisable at that time.

Nausea and vomiting

Nausea and vomiting are one of the main symptoms faced by pregnant women. The most important thing to help control nausea during pregnancy is to not allow the stomach to become empty. It is better to eat snacks at small intervals rather than eating three large meals in a day. During nausea, the woman will not want to eat, but if food is not consumed regularly, then blood sugar levels can drop and actually increase nausea during pregnancy. Other ways to avoid nausea and vomiting include:

1. drink a lot of water with a pinch of Celtic sea salt which will aid hydration and replace the lost electrolytes. Table salt is known to bloat, hence sea salt is recommended.
2. Vitamin B6 is known to treat nausea, and food items like beef, tuna, banana, chicken, liver, avocados and whole grains.
3. Ginger also will have the similar effect on nausea as vitamin B6, hence raw ginger chopped into food or blended with tea will help relieve nausea and vomiting.

Lack of rest and taking a lot of stress has negative impact on their bodies. It can increase nausea during pregnancy, a very relaxing massage and stress relieved mind will help them to control nausea. Sleep and rest is very important. Eating fruits and vegetables, could reduce pregnant women's risk of developing an upper respiratory tract infection (URTI).

Exercise and Diet

If they were into vigorous exercise before pregnancy and continue to do so during their pregnancy then they should take the same diet for 1st trimester and increase their diet intake by 300 calories for their second trimester. A known fact is that regular exercise during pregnancy can improve posture also help in decreasing discomforts such as backaches and fatigue. Physical activity may help in prevent Gestational diabetes, relieve stress, and build more stamina needed for labor and delivery. The heart rate should not exceed beyond 140 beats per minute while exercising, overdoing the exercises will increase the fatigue rather than cure it. Hence expectant mothers should not revert to their vigorous exercise schedule before pregnancy.

Second Trimester

Second trimester is a crucial stage in pregnancy, hence the diet should include all essential food groups that are carbohydrates, proteins, fats along with already consuming vitamins and minerals. Increase the consumption of fruits and vegetables to 4 - 5 servings of each, which is more compared to your first trimester. Many women suffer from constipation in their second trimester, hence fibre rich food helps in easing the bowel movement and relieves constipation. Try reducing fat intake in the diet plan, along with sugar, both will lead to increase in excess weight than required. Weight gain is unhealthy for pregnancy in this stage. Consume sufficient amount of dairy products that are rich in calcium, very healthy and essential mineral for the baby.

Third Trimester

Third trimester is the time when the baby is fully developed and the mother is getting ready for labour. Hence nutrient same as before should be continued, along with 300 calories extra per day to keep the mother energized. Drinking between the meals should be reduced as the pressure on bladder increases by the growing foetus inside the womb. Diet plan should be more or less same as the second trimester only increased quantities to suffice the exhaustion and tiredness face by the mother. Proteins can be found in fish also, but it is necessary to enquire from the doctor regarding which fish is best for the mother and baby, for example : Many ladies are advised against swordfish and salmon. For teeth and bone development of the baby skimmed milk, cheese and yoghurt should be consumed, as they are enriched with calcium. One glass of skimmed milk or low fat milk is a healthy option.

Edema

Most pregnant ladies suffer from swelling in their third trimester, a condition termed as edema. Consumption of sodium rich food items should be avoided. It is estimated about 75% of women experience it. The most effective ways to relieve from excess swelling and related discomforts is

- Avoid elastic topped socks or knee high stockings, try wearing support panty hose instead.
- Wear comfortable shoes, slip ons are the best type.
- Avoiding standing for long hours and try moving around. Keep your leg elevated whenever possible.
- Hydrotherapy; lying in a pool helps in relieving the body of excess fluids and still protecting the uterus.
- Excess or low amount of salt level in the body cause swelling, hence moderation is necessary.

Potentially harmful determinants during pregnancy

It is advised for pregnant women to pay special attention to food hygiene during pregnancy in addition to avoiding certain foods in order to reduce the risk of exposure to substances that may be harmful to the developing fetus.

This can include food pathogens and toxic food components, alcohol, and dietary supplements such as vitamin A. Dietary vitamin A is obtained in two forms which contain the preformed vitamin (retinol), that can be found in some animal products such as liver and fish liver oils, and as a vitamin A precursor in the form of carotenes, which can be found in many fruits and vegetables.

Intake of retinol, in extreme cases, has been linked to birth defects and abnormalities. However, regular intake of retinol is not seen as dangerous. It is noted that a 100 g serving of liver may contain a large amount of retinol, so it is best that it is not eaten daily during pregnancy, something which is also the same with alcohol intake in binge drinking.

Excessive amounts of alcohol have been proven to cause fetal alcohol syndrome. The World Health Organization recommends that alcohol should be avoided entirely during pregnancy, given the relatively unknown effects of even small amounts of alcohol during pregnancy.

Pregnant women are advised to pay particular attention to food hygiene and to avoid certain foods during pregnancy in order to minimize the risk of food poisoning from potentially harmful pathogens such as listeria, taxoplasmosis, and salmonella. Pregnant women are therefore advised to avoid foods in which high levels of the bacteria have been found, such as in soft cheeses. Listeria are destroyed by heat and therefore pregnant women are advised to reheat ready-prepared meals thoroughly.

Pregnant women should also wash their fruit and vegetables very thoroughly in order to minimize risk. Salmonella poisoning is most likely to come from raw eggs or undercooked poultry. Maternal obesity has a significant impact on maternal metabolism and offspring development.

Insulin resistance, glucose homeostasis, fat oxidation and amino acid synthesis are all disrupted by maternal obesity and contribute to adverse outcomes. Modification of lifestyle is an effective intervention strategy for improvement of maternal metabolism and the prevention of adverse outcomes.

PRECONCEPTION OF PREGNANCY

The health of a mother and baby is influenced not only by diet during pregnancy but also by diet before conception. Eating a healthy diet before pregnancy will give your body a good store of nutrients for the baby to draw on during pregnancy. The foetus is most susceptible to nutritional imbalance during the first few months of pregnancy because this is the time of most rapid development.

If the lady is very underweight or overweight, she should try to achieve an acceptable weight for her height by a sensible and well-balanced diet.

If she uses to take the contraceptive pill, it is a good idea to come off and use an alternative method of contraception a few months before conceiving. This is because the pill can alter levels of some nutrients, particularly vitamin B6, folate and zinc. Ensure that she have a well balanced diet with plenty of fresh fruit, vegetables (especially green vegetables) and wholegrain cereals. Try to avoid fatty foods, sweets, biscuits and cakes.

Now is the time to make other healthy lifestyle changes and so stop smoking, cut down on alcohol and drinks containing caffeine such as tea, coffee and cola. Try a barley drink such as Barley Cup, or herb and fruit teas instead.

If the lady find that she is pregnant before she have had time to think about preconception, then don't worry, there is still plenty of time to make healthy changes to the diet.

Pregnancy is a time when good nutrition is vital, for vegetarians and non-vegetarians alike. It is a time of readjustment as well as growth. The nine months are divided up into three divisions of three months each called trimesters. Many women experience changes in mood, activity and appetite with the different stages of pregnancy. There is no truth in the old saying that pregnancy means eating for two.

The extra energy needed is only 200-300 calories a day for nine months. This is equivalent to two slices of bread with margarine or a jacket potato with baked beans or cheese. Some women do feel a lot more hungry than this and if they are gaining weight at the right rate, they should eat according to their appetite.

0-3 Months

Women usually feel different and may experience tiredness as well as a feeling of sickness particularly in the morning. Calorie needs are only about 100kcal more in the early stages of pregnancy and some women do not experience much increase in appetite until the end of this period.

3-6 Months

Appetite usually increases after the first three months and this period requires about 300-400kcal extra calories a day. Extra calorie needs should be met by cereals, pulses, nuts and seeds, dairy products (unless vegan) and starchy vegetables such as potatoes.

6-9 Months

The baby is maturing now and this is a time for easing up on activity and preparing for the birth. The baby takes up a lot of space and may press on the stomach, reducing the capacity for food. Many women feel they need to eat small meals more frequently at this stage. Normal weight gain during pregnancy is about 22 to 28 pounds or one and a half to two stones (10-12.5kg). Weight gain often slows down during the last few months of pregnancy.

Exercise

It is a good idea to do some form of exercise during pregnancy. Antenatal clinics normally advise on appropriate exercise. Swimming or gentle yoga are often recommended.

Nutrients

Fluid

You may find that you are more thirsty during pregnancy. This is natural as fluid intake should increase. Never allow yourself to become over thirsty and include plenty of fresh water, dilute fruit juices, milk (soya or cow's) and herb teas. Drinks containing caffeine (tea, coffee and cola) should be limited and alcohol should be avoided altogether if possible.

Protein

Increased protein needs in pregnancy are usually met simply by the extra calories from more foods. Protein can be found in milk, cheese, eggs, soya milk, tofu, cereals, nuts and pulses. A normal variety of these foods will provide adequate protein.

Intake of dairy products and eggs should not increase dramatically. Some people believe that excessive amounts may sensitise the baby in the womb to allergies towards these foods.

Iron

The need for iron is increased during pregnancy, especially during the later stages. Anaemia, duc to iron deficiency, is common in pregnancy whether you are vegetarian or not. Vegetarians should be especially careful to include plenty of iron in their diet as vegetable sources are not as well absorbed.

Good vegetarian sources of iron can be found in wholegrain cereals, pulses, green vegetables and dried fruits. Iron absorption is increased if taken with a good source of vitamin C, which can be found in fresh fruit and vegetables. Tea contains tannin which can inhibit iron absorption and should not be taken an hour before or after a meal.

Many doctors routinely prescribe iron tablets for pregnant women. Iron tablets may not be needed unless a blood test demonstrates anaemia. Iron levels normally decrease during pregnancy as the blood becomes more dilute. Some women prefer to take a natural iron supplement, such as Floridix, which is available from health food stores.

Calcium and Vitamin D

The body needs extra calcium during pregnancy, especially in the later stages, to enable the baby's bones to develop. Calcium absorption from the gut is more efficient during pregnancy and this should provide enough to meet requirements. Vegans and vegetarian women who consume few dairy products need to be particularly careful to ensure adequate calcium in the diet.

Some vegan women, especially if they intend to breast-feed, may decide on a calcium supplement as a wise precaution, although with a good vegetable intake of calcium, it may not be necessary. Good sources of calcium include green vegetables, almonds, sesame seeds or tahini, cow's milk, tofu, cheese, yoghurt, wholegrain cereals and pulses. Vitamin D is essential for calcium absorption and can be obtained from sunlight, margarine and dairy products.

Folate

Folate is one of the B vitamins needed in increased amounts during pregnancy. Research has shown that a deficiency of folate during pregnancy can lead to birth defects. Vegetarians should not be at risk as the best sources of this vitamin are green leafy vegetables, fruit, peanuts, yeast extract and wholegrain cereals.

Vitamin B12

Vitamin B12 is essential to the growth and development of your baby. If adequate amounts of dairy products, eggs and fortified yeast extract are included in your diet, then you should have enough vitamin B12. It is especially important for vegans to include a reliable source of vitamin B12 in the diet during pregnancy. Some vegan foods, such as certain brands of soya milk, margarine and soya products are fortified with this vitamin. If the vitamin B12 in your diet is unreliable, then a supplement is recommended.

NUTRITION DURING LACTATION

A nutritional demand on the mother is great during lactation. The nutritional relationship between the mother and child continues for some more time after delivery. The newborn baby is given only breast milk for first 3 months of life so the mother's diet should be adequate with all needed nutrients.

Nutritional needs increased during lactation because

1. For producing sufficient breast milk for the infant.
2. For providing enough nutrients for the growing infant
3. For the mother's daily requirement.

3

Food and Nutrition Biochemistry

VITAMIN

A vitamin is an organic compound required as a nutrient in tiny amounts by an organism. The term vitamin was derived from "vitamine," a combination word from *vital* and *amine,* because it was suggested that the organic micronutrient food factors which prevented beriberi and perhaps other similar dietary-deficiency diseases, might be chemical amines. This proved incorrect for the micronutrient class, and the word was shortened. Today, a chemical compound is called a vitamin when it cannot be synthesized in sufficient quantities by an organism, and must be obtained from the diet. Thus, the term is conditional both on the circumstances and the particular organism. For example, ascorbic acid (vitamin C) is a vitamin for humans, but not for most other animals, and biotin and vitamin D are required in the human diet only in certain circumstances. The term *vitamin* does not include other essential nutrients such as dietary minerals, essential fatty acids, or essential amino acids, nor does it encompass the large number of other nutrients that promote health but are otherwise required less often.

Vitamins are classified by their biological and chemical activity, not their structure. Thus, each "vitamin" refers to a number of *vitamer* compounds that all show the biological activity associated with a particular vitamin. Such a set of chemicals are grouped under an alphabetized vitamin "generic descriptor" title, such as "vitamin A", which includes the compounds retinal, retinol, and four known carotenoids. Vitamers by definition are convertible to the active form of the vitamin in the body, and are sometimes inter-convertible to one another, as well.

Vitamins have diverse biochemical functions. Some have hormone-like functions as regulators of mineral metabolism (e.g. vitamin D), or regulators of cell and tissue growth and differentiation (e.g. some forms of vitamin A). Others function as antioxidants (e.g. vitamin E and sometimes vitamin C). The largest number of vitamins (e.g. B complex vitamins) function as precursors for enzyme cofactors, that help enzymes in their work as catalysts in metabolism. In this role, vitamins may be tightly bound to enzymes as part of prosthetic groups: for example, biotin is part of enzymes involved in making fatty acids. Alternately, vitamins may also be less tightly bound to enzyme catalysts as coenzymes, detachable molecules which function to carry chemical groups or electrons between molecules. For example, folic acid carries various forms of carbon group – methyl, formyl and methylene-in the cell. Although these roles in assisting enzyme-substrate reactions are vitamins' best-known function, the other vitamin functions are equally important.

Until the 1900s, vitamins were obtained solely through food intake, and changes in diet (which, for example, could occur during a particular growing season) can alter the types and amounts of vitamins ingested. Vitamins have been produced as commodity chemicals and made widely available as inexpensive pills for several decades, allowing supplementation of the dietary intake.

History

The value of eating a certain food to maintain health was recognized long before vitamins were identified. The ancient Egyptians knew that feeding liver to a patient would help cure night blindness, an illness now known to be caused by a vitamin A deficiency. The advancement of ocean voyage during the Renaissance resulted in prolonged periods without access to fresh fruits and vegetables, and made illnesses from vitamin deficiency common among ships' crews.

In 1749, the Scottish surgeon James Lind discovered that citrus foods helped prevent scurvy, a particularly deadly disease in which collagen is not properly formed, causing poor wound healing, bleeding of the gums, severe pain, and death. In 1753, Lind published his *Treatise on the Scurvy*, which recommended using lemons and limes to avoid scurvy, which was adopted by the British Royal Navy. This led to the nickname Limey for sailors of that organization. Lind's discovery, however, was not widely accepted by individuals in the Royal Navy's Arctic expeditions in the 19th century, where it was widely believed that scurvy could be prevented by practicing good hygiene, regular exercise, and by maintaining the morale of the crew while on board, rather than by a diet of fresh food. As a result, Arctic expeditions continued to be plagued by scurvy and other deficiency diseases. In the early 20th century, when Robert Falcon Scott made his two expeditions to the Antarctic, the prevailing medical theory was that scurvy was caused by "tainted" canned food.

During the late 18th and early 19th centuries, the use of deprivation studies allowed scientists to isolate and identify a number of vitamins. Initially, lipid from fish oil was used to cure rickets in rats, and the fat-soluble nutrient was called "antirachitic A". Thus, the first "vitamin" bioactivity ever isolated, which cured rickets, was initially called "vitamin A", although confusingly the bioactivity of this compound is now called vitamin D. In 1881, Russian surgeon Nikolai Lunin studied the effects of scurvy while at the University of Tartu in present-day Estonia. He fed mice an artificial mixture of all the separate constituents of milk known at that time, namely the proteins, fats, carbohydrates, and salts. The mice that received only the individual constituents died, while the mice fed by milk itself developed normally. He made a conclusion that "a natural food such as milk must therefore contain, besides these known principal ingredients, small quantities of unknown substances essential to life." However, his conclusions were rejected by other researchers when they were unable to reproduce his results. One difference was that he had used table sugar (sucrose), while other researchers had used milk sugar (lactose) that still contained small amounts of vitamin B.

The Ancient Egyptians knew that feeding a patient liver would help cure night blindness. In east Asia, where polished white rice was the common staple food of the middle class, beriberi resulting from lack of vitamin B1 was endemic. In 1884, Takaki Kanehiro, a British trained medical doctor of the Imperial Japanese Navy, observed that beriberi was endemic among low-ranking crew who often ate nothing but rice, but not among officers who consumed a Western-style diet. With the support of the Japanese navy, he experimented using crews of two battleships; one crew was fed only white rice, while the other was fed a diet of meat, fish, barley, rice, and beans.

The group that ate only white rice documented 161 crew members with beriberi and 25 deaths, while the latter group had only 14 cases of beriberi and no deaths. This convinced Takaki and the Japanese Navy that diet was the cause of beriberi, but mistakenly believed that sufficient amounts of protein prevented it. That diseases could result from some dietary deficiencies was further investigated by Christiaan Eijkman, who in 1897 discovered that feeding unpolished rice instead of the polished variety to chickens helped to prevent beriberi in the chickens.

The following year, Frederick Hopkins postulated that some foods contained "accessory factors"—in addition to proteins, carbohydrates, fats, et cetera—that were necessary for the functions of the human body. Hopkins and Eijkman were awarded the Nobel Prize for Physiology or Medicine in 1929 for their discovery of several vitamins.

In 1910, the first vitamin complex was isolated by Japanese scientist Umetaro Suzuki who succeeded in extracting a water-soluble complex of micronutrients from rice bran and named it aberic acid (later *Orizanin*). He published this discovery in a Japanese scientific journal.

When the article was translated into German, the translation failed to state that it was a newly discovered nutrient, a claim made in the original Japanese article, and hence his discovery failed to gain publicity. In 1912 Polish biochemist Kazimierz Funk isolated the same complex of micronutrients and proposed the complex be named "Vitamine" (a portmanteau of "vital amine"). The name soon became synonymous with Hopkins' "accessory factors", and by the time it was shown that not all vitamins were amines, the word was already ubiquitous. In 1920, Jack Cecil Drummond proposed that the final "e" be dropped to deemphasize the "amine" reference, after researchers began to suspect that not all "vitamines" (particularly vitamin A) had an amine component.

In 1931, Albert Szent-Györgyi and a fellow researcher Joseph Svirbely suspected that "hexuronic acid" was actually vitamin C, and gave a sample to Charles Glen King, who proved its anti-scorbutic activity in his long-established guinea pig scorbutic assay. In 1937, Szent-Györgyi was awarded the Nobel Prize in Physiology or Medicine for his discovery. In 1943 Edward Adelbert Doisy and Henrik Dam were awarded the Nobel Prize in Physiology or Medicine for their discovery of vitamin K and its chemical structure. In 1967, George Wald was awarded the Nobel Prize (along with Ragnar Granit and Haldan Keffer Hartline) for his discovery that vitamin A could participate directly in a physiological process.

In Humans

Vitamins are classified as either water-soluble or fat soluble. In humans there are 13 vitamins: 4 fat-soluble (A, D, E and K) and 9 water-soluble (8 B vitamins and vitamin C). Water-soluble vitamins dissolve easily in water, and in general, are readily excreted from the body, to the degree that urinary output is a strong predictor of vitamin consumption. Because they are not readily stored, consistent daily intake is important. Many types of water-soluble vitamins are synthesized by bacteria. Fat-soluble vitamins are absorbed through the intestinal tract with the help of lipids (fats). Because they are more likely to accumulate in the body, they are more likely to lead to hypervitaminosis than are water-soluble vitamins. Fat-soluble vitamin regulation is of particular significance in cystic fibrosis.

MINERALS

Minerals are elementary chemical compounds whose atomic characteristics make them indispensable to the organism. Minerals include trace elements and are essential because they cannot be produced by the organism and therefore must be found, through food and drinking water in the mineral environment; this depends on the nature of the soil and the surface water. Minerals include major elements such as calcium and phosphorus, and trace elements of which the organism contains less than a few grams (sometimes as little as a few milligrams).

Although some minerals are present in the organism in only minute amounts, this does not mean that their deficiency is harmless. Mineral biochemistry is a developing science, and it is far from mastered. As a result, detailed information is not available for all essential minerals.

The main minerals are the electrolytes (sodium, potassium, and chlorine), the minerals found mainly in the skeleton (calcium, phosphorus, and magnesium), and iron, zinc, iodine, copper, selenium, manganese, molybdenum, and chromium. Cobalt could also be mentioned, but it seems that it only acts as a constituent of vitamin B_{12}. Trace metals exist, such as tin, bromine, lead and probably others that may well be essential to man, but research has yet to confi rm this.

Specifi c defi ciencies have not been identifi ed either. In terms of nutritional needs, electrolytes do not pose a major problem, other than in cases of dehydration. Phosphorus is so abundant in the vegetable kingdom that dietary deficiency is virtually impossible. It is also likely that calcium deficiency due to dietary inadequacy is rare, and problems of osteomalacia and rickets are due to vitamin D deficiency, and osteoporosis to demineralization, whose possible connection to an inadequate dietary calcium intake remains unknown.

Sulphur is supplied by proteins: its deficiency is therefore likely to be concealed by protein deficiency, or its effects to be confused with the latter's.

Copper, manganese, magnesium, chromium, molybdenum and selenium defi ciencies do not pose public health problems, because the diet should in principle provide adequate supplies. However, in the case of severe malnutrition causing the complete loss of tissue components, or of seriously restricted access to food (as is the case in some imprisonment situations), deficiency in these minerals and others should be taken very seriously. They must then be administered in therapeutic doses when treating severe malnutrition, and at levels equivalent to reference intakes in the case of imprisonment or any other situation where the diet is known to be defi cient. Ultimately, the minerals whose clinical deficiency is most frequently confi rmed are iron and iodine. It is however possible that sub-clinical zinc deficiency is more common, and plays a greater role than thought until recently in retarded growth and loss of appetite, which would corroborate zinc belonging to Type II nutrients.

Electrolytes

The three main electrolytes are sodium, potassium and chlorine. They account for the osmotic pressure of fl uids in the organism, and their repartition in the body determines their volume within (intra-cellular) and outside (extra-cellular) cells.

- Sodium is specifi cally an extra-cellular cation (or positively-charged ion), whose role is to maintain the volume of extra-cellular fl uids, the osmotic pressure of these fl uids, and the acid-base balance. It also contributes to the transmission of impulses in and between nerves and muscle, and to the transport phenomena of metabolites within cells.
- Unlike sodium, potassium is a specifi cally intra-cellular cation, together with magnesium, organic phosphates and protein anions, where its role is complementary to that played by sodium, for functions that are essentially the same as those of sodium.
- Chlorine is the electrochemical counterpart of sodium and potassium.

The chemistry of electrolytes is quite complex, as is that of hydrous and acid-base balances. These aspects do not relate to humanitarian action because when disorders arise, they require medical expertise and laboratory facilities. In crisis settings, on the other hand, victims (malnourished or not) often require rehydration salts whose composition differs according to whether there is severe malnutrition or not. Dehydration can have two causes.

1. Water shortage that manifests itself in increased plasma chlorine and sodium levels; this occurs in case of insuffi cient ingestion of water, either because its availability is insuffi cient, or because one does not receive enough. This is typical when one depends on third parties to obtain drinking water, as is the case for small children but also, frequently, prisoners. The treatment simply consists of drinking water or, as a last resort only, the intravenous administration of a 2.5–5% dextrose solution (pure water would cause haemolysis).
2. The combined deficiency of chlorine, sodium and water, the most frequent form of dehydration by far and whose most common – and thus best-known – causes are diarrhoea, vomiting and excessive perspiration. The preferred treatment of diarrhoea is the oral administration of rehydration salts dissolved in water; these salts are found in sachets and are based on two standard formulae:

either	*or*
3.5 g of sodium chloride	3.5 g of sodium chloride
2.5 g of sodium bicarbonate	2.9 g of trisodium citrate dihydrate
1.5 g of potassium chloride	1.5 g of potassium chloride
20 g of glucose	20 g of glucose

Both formulae are to be dissolved in 1 litre of drinking water. The dosage is 1 litre per 24 hours for infants, 1 litre per 8 to 24 hours (according to age) for children, and at will for adults. The primary purpose is rehydration, and then the compensation of losses as they occur.

Other specifi c defi ciencies, such as chlorine or potassium deficiency, result mainly from medical problems (thus warranting medical expertise), and rarely arise in humanitarian action.

Electrolyte surges are essentially caused by the overly high intake of salt or by physiological disorders. It has long been thought that excessive sodium intake *via* salt affected high blood pressure signifi cantly. This view was later seriously challenged, but nowadays it does appear that salt intake greater than 8 g/day does indeed result in an increase in blood pressure. It is therefore recommended not to exceed this intake, which must allow for the diet's salt content. Moreover, excessive salt intake displaces calcium from the bones and could thus contribute to osteoporosis.

Bone minerals

The three minerals predominant in the bones are calcium, bound in the bones to phosphorus (in a crystal form called hydroxyapatite), and magnesium.

Calcium

Source: calcium is found (in decreasing order of concentration) in hard cheese (750 mg/100 g), oilseeds and in particular almonds, walnuts and hazelnuts (175 mg/100 g), soft cheese and milk (130 mg/100 g), and fi nally dried fruit (100 mg/100 g) (Randoin, 1982).

Absorption: calcium is absorbed in two ways. One is active, and is regulated by vitamin D, mainly in the duodenum; the other is passive and occurs throughout the intestine, but mainly in the ileum. Calcium absorption also depends largely on the other nutrients contained in the diet, hormonal levels, and physiological conditions such as pregnancy and breastfeeding. The absorption rate of calcium is only 25 to 50% of the ingested amount, and the remainder is excreted in the faeces. However, in case of need, the organism can retain much more calcium – as a result, deficiency resulting from an inadequate diet is rare.

Metabolism, storage and excretion: bones are living tissue, undergoing changes permanently. The calcium contained in the organism therefore perpetually circulates between the bones (that contain 99% of the organism's overall calcium) and the bloodstream that contains the remaining 1%. The concentration of calcium in the bloodstream is strictly regulated because of the vital role played by calcium in many physiological processes. This regulation occurs through absorption, excretion and calcium circulation between the bloodstream and the bones, which store it. The parathyroid hormone, calcitonin (secretions of the parathyroid and thyroid gland respectively) and vitamin D all regulate the homeostasis of calcium. In the bones, the calcium metabolism is determined by age, hormonal secretion, and physical exercise. The skeleton demineralizes in postmenopausal women and men above 65; in women, the impact is far greater because of the hormonal changes associated with menopause. Physical exercise maintains the bony mass, which atrophies in case of immobility just like any unused organ. Calcium is excreted in practically equal parts in the urine and the faeces.

Function: calcium plays a fundamental structural role in bones and teeth, and is their chief mineral constituent. Calcium also performs a fundamental trigger function in a number of physiological processes such as enzyme activation, muscle contraction, vesicular secretion, the transmission of nerve impulses, the clotting of blood and cell division.

Deficiency: calcium deficiency caused by dietary inadequacy is rare, and is limited to specifi c circumstances (e.g. the overall defi cient diet in some prison settings). In the case of osteoporosis (the thinning of bones observed in old age), the administration of calcium supplements does not appear to impede decalcifi cation signifi cantly. As an illustration, in developing countries where calcium intake is minimal but physical exercise considerable, osteoporosis is less common in postmenopausal women than in industrialized countries, where calcium intake is greater, but physical exercise is limited. The preferred treatment of osteoporosis is the administration of oestrogen. In most cases, intake is adequate in view of the adaptability of absorption in case of low intake. Breastfeeding women secrete constant amounts of calcium in maternal milk, whatever their intake. This indicates that an intake that is inadequate in terms of breastfeeding is compensated by the calcium stored in the bones, implying a demineralization of the bones. Nevertheless, this demineralization is reversible after the cessation of breastfeeding.

Requirements: calcium requirements are determined by excretion and absorption effi ciency, but the latter itself is infl uenced by a number of factors such as diet. As a result, it is diffi cult to formulate recommendations. On the other hand, additional needs related to breastfeeding are known and should be incorporated into the diet to avert the risk of bone demineralization in the mother.

Pharmacological intake: pharmacological intakes of calcium appear to be useless, and recommended intakes are therefore suffi cient.

Toxicity: overly high calcium intake is not toxic, and does not increase the risk of kidney stone development. It does however impede iron absorption, and it is therefore preferable to adhere to recommended intakes.

Phosphorus

Source: phosphorus is abundant in animal products, oilseeds, legumes and cereals. It is therefore improbable for its deficiency to be caused by an inadequate diet.

Absorption: man absorbs between 60 and 70% of ingested phosphorus, and the absorption of animal phosphorus is more effi cient than that of vegetable origin. Like calcium, the smaller the amount of phosphorus supplied by the diet, the more effi cient its absorption is. To date, phosphorus absorption remains ill-known; however, both an active, saturable, transport mechanism and a non-saturable mechanism (depending on the phosphorus concentration) exist. Vitamin D, which facilitates the absorption of calcium, contributes to the absorption of phosphorus likewise.

Metabolism, storage and excretion: the phosphorus homeostasis occurs in the intestine, the kidneys, bone and lean mass. Plasma phosphorus levels are mainly regulated by the kidneys that reabsorb or excrete phosphorus according to intake. The intestine, however, also plays a signifi cant role by improving its absorption effi ciency in case of low intake. Like calcium, phosphorus is mainly stored in the bones and teeth (that together contain 85% of the overall phosphorus in the organism), while 14% is found in muscle. Phosphorus excretion occurs fi rstly in the kidney, and its re-absorption is regulated by the parathyroid hormone, according to intake.

Function: phosphorus performs a number of essential functions in the organism. It combines with calcium in bone formation, it serves as a construction unit in the great molecular formations of living beings, it is indispensable to the storage and release of chemical energy, it is essential to membrane structure and permeability, and it regulates many enzyme activities.

Deficiency: phosphorus deficiency caused by dietary inadequacy is rare. However, in the elderly, phosphorus absorption falls, whereas its excretion in the urine increases; this can cause a deficiency that manifests itself by anorexia, muscle weakness, debility and bone pains. Requirements: the phosphorus needs are not well known, and they are therefore assimilated on a molar basis to calcium needs.

Toxicity: infants fed on unbalanced formula supplying excessive phosphorus with respect to calcium (i.e. more than 2 parts of phosphorus for 1 part of calcium) can suffer hypocalcaemia and tetany. Recommended intakes should therefore be adhered to strictly, and the phosphorus-calcium ratio should be inferior to 2 (>2:1).

Magnesium

Source: all unprocessed foods contain magnesium, but concentrations vary greatly. Milk, wholemeal cereals, legumes, vegetables and potatoes contain particularly high concentrations of magnesium. Vegetables are generally rich in magnesium (its role to chlorophyll is analogous to that of iron to haemoglobin).

Absorption: the absorption of magnesium, as is the case for other minerals, occurs through an active mechanism for moderate intakes, and this mechanism saturates as intakes increase. Magnesium is then absorbed passively. Absorption rates vary between 20 and 70% of the intake.

Metabolism, storage and excretion: approximately two-thirds of magnesium are found in the bones and teeth, one-quarter in muscle, and practically all the rest in other cells, while less than 1% is found in extra-cellular fl uids.

Magnesium's homeostasis is similar to calcium's and occurs between the reserves stored in the bones, intestinal absorption and urinary excretion. Little is yet known regarding its hormonal regulation mechanism. Its excretion through urine is as adaptable as its intestinal absorption.

Function: magnesium is a vital element and performs many, similarly important, functions.

More than 300 reactions in the organism are currently attributed to magnesium. It is of particular importance to the building and development of the bone structure, the synthesis of lipids and proteins, phosphorylation and de-phosphorylation mechanisms, metabolic energy production channels and the maintenance of normal nerve signalling and muscle contraction.

Deficiency: magnesium deficiency caused by dietary inadequacy is improbable because all unprocessed foods, including staple foods, generally provide a good source of magnesium, and the average daily intake is usually appropriate (150 to 500 mg).

However, deficiency can result from malabsorption and during recovery from severe malnutrition. It is therefore fundamental to ensure an adequate magnesium intake during nutritional catch-up, because it is a Type II nutrient. Requirements: because dietary inadequacy can hardly account for magnesium deficiency, it is diffi cult to determine magnesium requirements accurately – reference to a suitable intake range (150 to 500 mg/day) is therefore preferable. On the other hand, magnesium requirements during recovery from severe malnutrition are quite precise.

Toxicity: overly high oral intake is not toxic because of the adaptability of magnesium excretion by the kidney. Excessive intake can nevertheless cause passing diarrhoea. On the other hand, excessive intravenous administration can cause a fatal depression of the central nervous system and muscular paralysis.

Other minerals

Iron

Source: the following, listed in order of importance, are sources of iron (Randoin, 1982):

- parsley (> 10 mg/100 g): although excellent, it is uncommon to eat parsley as a vegetable, and iron intake through parsley is therefore usually limited;
- liver, legumes and egg yolk all contain between 5 and 10 mg/100 g; egg yolk unfortunately inhibits iron absorption (that is, its own iron content, and that supplied by other foods);
- meat, oilseeds, wholemeal fl our, watercress and spinach all contain between 2 and 4 mg/100 g;
- most fresh fruits and vegetables, cheese and white fl our all contain between 1 and 0.5 mg/100 g; they are thus poor in iron, like fermented beverages, milk and fruits such as apples, pears and cherries that are poorer still, containing less than 0.4 mg/100 g;
- the iron contained in drinking water and in cooking utensils can represent a signifi cant source, but it is diffi cult to take them into account owing to their considerable variability.

Absorption: iron absorption is a phenomenon of particular importance, because it is at this level that the organism attempts to maintain a balance so as to avoid deficiency and excess, without resorting to loss modulation. Iron absorption occurs in two ways, according to whether it is bound to the haemoglobin or myoglobin haeme group. Isolated non-haeme iron is dissolved by gastric juices, reduced into Type II iron (i.e. ferrous iron), and bound to other molecules. Where such molecules are small – such as ascorbic acid, sugars, and amino acids – they promote the absorption of iron. Some elements nevertheless impede its absorption; these are tannins, phytates, phosphates (found in wholemeal cereals and oilseeds), egg yolk, tea, and coffee. Milk and soy proteins also inhibit iron absorption. The phytates and tannins contained in tea and coffee can reduce iron absorption by approximately half. Absorption proper occurs in the small intestine, particularly in the duodenum. Organic acids, bile salts, and the peptide by-products of protein digestion all facilitate passage through the mucous membrane.

Iron binds with a specifi c protein to penetrate epithelial cells. Upon entering plasma, iron is oxidized into Type III iron (ferric iron), most is then bound to transferrin, and the rest to ferritin. Haeme iron obtained from meat/fi sh and blood is absorbed in a different way. The haeme group is absorbed directly by the cells of the intestinal mucous membrane; the iron is then separated from the haeme group and delivered into the plasma in the form of Type III iron. The absorption of haeme iron is considerably higher (approximately 25%) than that of non-haeme iron (approximately 10 to 15%); it is not affected by absorption inhibiting or promoting factors that affect non-haeme iron. Even in small amounts, haeme iron contained in the diet considerably increases the absorption of non-haeme iron. Generally speaking, this comment applies to the different promoters and inhibitors of iron absorption whose effects are cumulative. As an example, the ingestion of a small portion of fi sh or meat together with cereals greatly increases the absorption of the iron contained in the latter; this effect will be strengthened by the consumption of fruit, or weakened by the consumption of tea. Calcium, for its part, inhibits iron absorption (both haeme and non-haeme); a glass of milk containing approximately 160 mg of calcium inhibits iron absorption by more than 50%.

Iron absorption increases with food intake, but in decreasing proportions. The mechanisms regulating iron absorption boost it in case of iron deficiency, and restrict it in case of excess – the reasons for this remain unclear. Healthy individuals absorb approximately 5 to 10% of the iron contained in their food, whereas defi cient individuals absorb between 10 and 20%. From a quantitative perspective, the maximum iron absorption among healthy adults is of the order of 1 to 2 mg/day, while it is in the vicinity of 3 to 6 mg for defi cient individuals (Shils, 1994).

Metabolism, storage and excretion: after its absorption, iron is transported to the bone marrow to be integrated into the haemoglobin during the synthesis of the red blood cells. From there, it remains in the blood together with the red cells for approximately 4 months. Following the destruction of the red blood cells by phagocytes (the lifespan of red blood cell is approximately 4 months), most of the iron returns to the plasma and extra-cellular fl uids, and a small amount is stored with ferritin and haemosiderin, which return their iron to plasma after their destruction, according to their turnover cycle. Most of the iron then resumes its cycle by returning to the bone marrow for the synthesis of red blood cells. The iron transport protein in the plasma is transferrin; it transports the iron to the bone marrow, but also to any other cell that requires it. Approximately two-thirds of the overall iron contained in the organism is located in the haemoglobin, approximately one-quarter is stored, a small proportion is found in a muscle protein (myoglobin), and the rest is found in extra-cellular fl uids and cells.

Less than 0.1% of the iron is in circulation with transferrin. The organism is remarkably effi cient in saving iron through almost complete recycling, whatever its level within its cycle between plasma and red blood cells. Deficiency is therefore mainly caused by abnormal losses and limited absorption. The excretion of iron is largely passive and related to inevitable losses or "leaks" of the organism: through the intestine, urine, bile, exfoliation (shedding of the skin) – together accounting for approximately 1 mg/day – and menstrual bleeding in women of childbearing age (in the vicinity of 25 mg per cycle). Pregnancy and breastfeeding also demand high iron levels in comparison to the absorption capacity: beyond normal losses that persist throughout pregnancy, 200 to 370 mg are transferred to the foetus, some 90 mg are utilized for the placenta and the umbilical cord, and an average 250 mg are lost in the bleeding associated with delivery; breastfeeding provides the infant with 0.5 to 1 mg of iron per day.

Function: the primary function of iron is obviously to transport oxygen and carbon dioxide in the organism, *via* red blood cells. Iron is vital for the production of anaerobic energy, being present in the active groups of most of the enzymes that contribute to this production. It also contributes to the transport of electrons by cytochromes in the respiratory chain.

Deficiency: iron deficiency is no doubt the most common deficiency, and mainly affects women of childbearing age. It is discussed below.

Requirements: a distinction must be made between absorbed and ingested iron, because the amount absorbed is considerably lower than that contained in the diet.

Pharmacological intake: there is no justifi cation for pharmacological iron intakes, particularly in view of its toxicity. However, supplements are often necessary to correct deficiency.

Toxicity: the excessive intake of iron is toxic. A normal diet, however, does not lead to iron excesses, because its absorption is regulated effi ciently. Excess pathologies – haemachromatosis (where excess iron deposited in organs causes them damage) or haemosiderosis (iron overload without obvious tissue damage) – can arise only if regulation is inadequate (hereditary metabolic aberrations), in case of excessive intake resulting from specifi c circumstances, or in the case of particular ailments, such as hepatic cirrhosis, and pancreatic insuffi ciency. The circumstances leading to excessive iron intake include protracted consumption of supplements by individuals who are not anaemic, and the consumption of traditional beverages fermented in iron containers that oxidize and release large amounts of iron into the fermentation medium.

Iodine

Source: the iodine content of food is mainly determined by the iodine concentration in the soil and groundwater, and this concentration is related to geology. During the formation of the earth, iodine was quite widespread, but was subsequently leached by rain, erosion, glaciations and fl oods. As a result, most terrestrial iodine is now found in the seas and in deep terrestrial layers, in the chemical form of iodide. Solar radiation transforms marine iodine into elementary iodine, which is volatile and enters the atmosphere before returning to earth through precipitations; it fi nally returns to the sea according to rainfall and drainage.

This iodine cycle nevertheless does not compensate for losses over time, as iodine returning to earth is instantly evacuated by soil drainage. This is why rugged mountains, well-irrigated and well-drained highlands, and fl ood-prone plains are most vulnerable to disorders related to iodine deficiency.

It is not useful for tables to indicate the iodine content of different food types, because of its high variability. On the other hand sea fi sh, seafood and seaweed are iodine-rich, as are coastal plants and the animals that eat them. Nowadays the most reliable source of iodine in many countries is salt, enriched in potassium iodate at between 30 to 50 mg of iodine/kg of salt (PAHO, 1986) – this is referred to as iodized salt. The iodine must be in the form of potassium iodate and not potassium iodide, which is too vulnerable to humidity and heat. One gram of potassium iodate contains 0.6 g of iodine. Thus, enrichment will consist of 50 to 83 mg of potassium iodate/kg of salt.

Function: iodine is an essential element used by the thyroid gland to produce thyroid hormones; the latter act as regulators of the metabolism, and ensure normal growth and development (both physical and mental) in children.

Deficiency: iodine deficiency is a serious threat to approximately one-quarter of humanity and, in some regions, constitutes a major public health problem.

Zinc

Source: zinc is found in most foods, but its concentration varies greatly. Its major sources include animal products (3–5 mg/100 g), cereals and legumes (2–3 mg/100 g). In countries where the consumption of animal products is high, the diet contains an adequate amount of zinc. In contrast, where animal products are not a major component of the diet, cereals and legumes provide most of the required zinc; this can prove inadequate if the cereals are overly processed or refi ned, because zinc is located mainly in the peripheral layers of the grain, and these are shed proportionately to the degree of processing. For example, wholemeal wheat fl our contains approximately 5.5 mg of zinc/ 100 g, whereas white fl our only contains 1.7 mg/100 g i.e. a 70% loss.

Absorption: zinc absorption is similar to that of iron insofar as it increases with its intake, but in decreasing proportions. Zinc is absorbed in the small intestine. The level of its absorption depends on the presence of promoting or inhibiting molecules; the latter are phytates, tannins and oxalates. On the other hand, zinc inhibits the absorption of copper.

Metabolism, storage and excretion: the zinc content in the organism depends on the effi ciency of its absorption, which rises in case of shortage or deficiency, and falls in case of excess. However, as mentioned above, absorption continues to increase according to intake, even if in a smaller proportion; excesses must therefore be disposed of, because zinc can be toxic. This is done through faecal excretion, which increases in case of excessive intake. Zinc absorption increases in the fi nal stages of pregnancy; it decreases with age, but so do losses, and this facilitates the maintenance of a balance. Zinc is not stored, strictly speaking, and insuffi cient intakes rapidly lead to indications of deficiency.

Function: zinc performs many essential functions in the organism, and is included in Type II nutrients accordingly. It is necessary to no less than 200 enzymes, it is a constituent of the cell membrane, and contributes to the stabilization of ribonucleic and desoxyribonucleic acids (RNA and DNA respectively) and ribosomes. The considerable contemporary research surrounding zinc may well lead to the discovery of further functions.

Deficiency: acute zinc deficiency is caused by a genetic disease – known as *acrodermatitis enteropathica* and characterized by an impaired zinc uptake from the intestine, zinc-defi cient parenteral nutrition (i.e. intravenous), different types of malabsorption, alcoholism, and kidney and metabolic disorders. In the Near East, zinc deficiency is blamed for dwarfi sm among young teenagers and delayed sexual maturity. Moderate zinc deficiency could be more common than assumed until recently, and results in growth retardation. Zinc deficiency develops with the severity of acute malnutrition, and addressing this condition requires the administration of zinc in therapeutic doses, because zinc determines nutritional recovery as much as other Type II nutrients do.

Toxicity: excessive zinc intake can cause chronic or acute disorders. Acute intoxication causes gastric pain, nausea, vomiting, and diarrhoea. Intoxication can result from the ingestion of beverages contaminated by the release of zinc in galvanized containers. Doses above 200 mg always cause vomiting. Beyond this threshold, they can be fatal. Chronic intoxication begins with zinc supplementation doses no higher than 25 mg/day, which result in a secondary copper deficiency by competition between the two elements in their intestinal absorption. Zinc supplements of the order of 150 mg/day result in low plasma levels of high-density lipoproteins (the risk of cardio-vascular disease is inversely proportional to these levels), and gastric erosion, and weaken the immune system (Shils, 1994).

Copper

Source: most foods contain copper, the richest sources (0.3–2 mg/100 g) being shellfi sh, oilseeds (including cocoa powder), legumes, and wholemeal cereals.

Absorption: like other metals, the absorption of copper increases with intake, but in decreasing proportions. Zinc inhibits the absorption of copper.

Metabolism, storage and excretion: following its absorption, copper is transported to the liver where it binds with a protein called ceruloplasmin, which transports it to the tissues. It is possible that copper could be stored in the liver, but in limited amounts. Little is known about the copper excretion mechanism, other than in the intestine by bile salts. The greater the copper intake or deficiency, the smaller its excretion is. Copper homeostasis depends on its intake and excretion.

Function: copper contributes to a number of oxidation-reduction reactions, as an enzyme constituent. These enzymes are essential in the respiratory chain, the synthesis of collagen proteins, the synthesis and maintenance of myelin, the synthesis of neuro-transmitters, the iron metabolism, and the protection against oxidation risks. Like zinc, copper is assumed to perform many more functions and these are currently under investigation.

Deficiency: copper deficiency can affect preterm babies, infants fed on non-modifi ed cow milk, and children recovering from severe malnutrition. Copper deficiency manifests itself by leucopaenia, neutropaenia, skeletal anomalies, and higher liability to infection. It is possible that inadequate intakes cause cardio-vascular disease in adults.

Toxicity: copper is toxic and intoxication is accidental. It is caused by the ingestion of acid beverages stored in copper containers or by winegrowers who use copper sulphate. Symptoms of copper intoxication include gastric pain, nausea, vomiting, and diarrhoea. Acute intoxication leads to coma, hepatic necrosis, vascular collapse and, ultimately, death.

Low white blood cell levels in the blood.

Low levels of white blood cells known as neutrophils in the blood.

Selenium

Source: selenium is found mainly in sea products, offal, meat, cereals, and legumes.

Absorption: selenium is usually ingested together with methionine and cysteine, and is well absorbed. Selenium homeostasis does not depend on its absorption.

Metabolism, storage and excretion: little is known about the selenium metabolism, but its excretion is mainly urinary, and its regulation determines selenium homeostasis.

Function: the main function of selenium is to act as an antioxidant agent. It probably performs other functions, such as the protection of cell immunity, and these are currently under investigation.

Deficiency: selenium deficiency causes Keshan disease, a fatal form of cardiomyopathy (heart muscle disease); however, the involvement of other factors appears to be required for the disease to develop. It is possible that selenium deficiency plays a role in the aetiology (i.e. cause) of kwashiorkor.

Toxicity: selenium overdose can originate in the diet in areas where the soil is selenium-rich, or by the ingestion of tablets. High doses of selenium are toxic, and cause nausea, diarrhoea, irritability, fatigue, peripheral neuropathy, balding, and brittle nails (Shils, 1994).

Chrome

Chrome is today recognized to be an essential nutrient that promotes the action of insulin and thus infl uences the mechanisms of glucides, lipids and proteins (Shils, 1994). Human deficiency has not been demonstrated to date.

Other trace metals

Other trace metals appear to be essential and their deficiency poses no particular problem because their intake is usually adequate. These include boron, which infl uences the mineral metabolism, manganese, an enzyme activator and a constituent of some enzymes, molybdenum, also a constituent of some enzymes, and silicon, which contributes to the synthesis of bones and cartilage.

Vanadium, bromine, fluorine, lead and tin could also be essential trace metals, but this remains to be proven. On the other hand, pharmacological doses of fl uorine provide protection against dental cavities, and possibly against fractures associated with osteoporosis.

WATER

All foods contain water. Vegetables in their natural condition contain large amounts, often 95 per cent, while in meats there is from 40 to 60 per cent or more. Prepared cereal products, as flour, corn meal, and oatmeal, which are apparently dry, have from 7 to 14 per cent. In general the amount of water in a food varies with the mechanical structure and the conditions under which it has been prepared, and is an important factor in estimating the value, as the nutrients are often greatly decreased because of large amounts of water. The water in substances as flour and meal is mechanically held in combination with the fine particles and varies with the moisture content, or hydroscopicity, of the air. Oftentimes foods gain or lose water to such an extent as to affect their weight; for example, one hundred pounds of flour containing 12 per cent of water may be reduced in weight three pounds or more when stored in a dry place, or there may be an increase in weight from being stored in a damp place. In tables of analyses the results, unless otherwise stated, are usually given on the basis of the original material, or the dry substance. Potatoes, for example, contain 2½ per cent of crude protein on the basis of 75 per cent of water; or on a dry matter basis, that is, when the water is entirely eliminated, there is 10 per cent of protein.

The water of foods is determined by drying the weighed material in a water or air oven at a temperature of about 100° C, until all of the moisture has been expelled in the form of steam, leaving the dry matter or material free from water. The determination of dry matter, while theoretically a simple process, is attended with many difficulties. Substances which contain much fat may undergo oxidation during drying; volatile compounds, as essential oils, are expelled along with the moisture; and other changes may occur affecting the accuracy of the work. The last traces of moisture are removed with difficulty from a substance, being mechanically retained by the particles with great tenacity. When very accurate dry matter determinations are desired, the substance is dried in a vacuum oven, or in a desiccator over sulphuric acid, or in an atmosphere of some non-oxidizing gas, as hydrogen.

Dry Matter

The dry matter of a food is a mechanical mixture of the various compounds, as starch, sugar, fat, protein, cellulose, and mineral matter, and is obtained by drying the material. Succulent vegetable foods with 95 per cent of water contain only 5 per cent of dry matter, while in flour with 12 per cent of water there is 88 percent, and in sugar 99 per cent. The dry matter is obtained by subtracting the per cent of water from 100, and in foods it varies from 5 per cent and less in some vegetables to 99 per cent in sugar:

1, Desiccator;

2, muffle furnace for combustion of foods and obtaining ash;

3, water oven for drying food materials.

Ash

The ash, or mineral matter, is that portion obtained by burning or igniting the dry matter at the lowest temperature necessary for complete combustion. The ash in vegetable foods ranges from 2 to 5 per cent and, together with the nitrogen, represents what was taken from the soil during growth. In animal bodies, the ash is present mainly in the bones, but there is also an appreciable amount, one per cent or more, in all the tissues. Ash is exceedingly variable in composition, being composed of the various salts of potassium, sodium, calcium, magnesium, and iron, as sulphates, phosphates, chlorides, and silicates of these elements. There are also other elements in small amounts. In the plant economy these elements take an essential part and are requisite for the formation of plant tissue and the production in the leaves of the organic compounds which later are stored up in the seeds. Some of the elements appear to be more necessary than others, and whenever withheld plant growth is restricted. The elements most essential for plant growth are potassium, calcium, magnesium, iron, phosphorus, and sulphur.

In the animal body minerals are derived, either directly or indirectly, from the vegetable foods consumed. The part which each of the mineral elements takes in animal nutrition is not well understood. Some of the elements, as phosphorus and sulphur, are in organic combination with the nitrogenous compounds, as the nucleated albuminoids, which are very essential for animal life. In both plant and animal bodies, the mineral matter is present as mineral salts and organic combinations. It is held that the ash elements which are in organic combination are the forms mainly utilized for tissue construction. While it is not known just what part all the mineral elements take in animal nutrition, experiments show that in all ordinary mixed rations the amount of the different mineral elements is in excess of the demands of the body, and it is only in rare instances, as in cases of restricted diet, or convalescence from some disease, that special attention need be given to increasing the mineral content of the ration. An excess of mineral matter in foods is equally as objectionable as a scant amount, elimination of the excess entailing additional work on the body.

The composition of the ash of different food materials varies widely, both in amount, and form of the individual elements. When for any reason it is necessary to increase the phosphates in a ration, milk and eggs do this to a greater extent than almost any other foods. Common salt, or sodium chloride, is one of the most essential of the mineral constituents of the body. It is necessary for giving the blood its normal composition, furnishing acid and basic constituents for the production of the digestive fluids, and for the nutrition of the cells. While salt is a necessary food, in large amounts, as when the attempt is made to use sea water as a beverage, it acts as a poison, suggesting that a material may be both a food and a poison. When sodium chloride is entirely withheld from an animal, death from salt starvation ensues. Many foods contain naturally small amounts of sodium chloride.

Organic Matter

That portion of a food material which is converted into gaseous or volatile products during combustion is called the organic matter. It is a mechanical mixture of compounds made up of carbon, hydrogen, oxygen, nitrogen, and sulphur, and is composed of various individual organic compounds, as cellulose, starch, sugar, albumin, and fat. The amount in a food is determined by subtracting the ash and water from 100.

The organic matter varies widely in composition; in some foods it is largely starch, as in potatoes and rice, while in others, as forage crops consumed by animals, cellulose predominates.

The nature of the prevailing organic compound, as sugar or starch, determines the nutritive value of a food. Each has a definite chemical composition capable of being expressed by a formula.

Considered collectively, the organic compounds are termed organic matter. When burned, the organic compounds are converted into gases, the carbon uniting with the oxygen of the air to form carbon dioxide, hydrogen to form water, sulphur to form sulphur dioxide, and the nitrogen to form oxides of nitrogen and ammonia.

Classification of Organic Compounds

All food materials are composed of a large number of organic compounds. For purposes of study these are divided into classes. The element nitrogen is taken as the basis of the division. Compounds which contain this element are called nitrogenous, while those from which it is absent are called non-nitrogenous. The nitrogenous organic compounds are composed of the elements nitrogen, hydrogen, carbon, oxygen, and sulphur, while the non-nitrogenous compounds are composed of carbon, hydrogen, and oxygen. In vegetable foods the non-nitrogenous compounds predominate, there being usually from six to twelve parts of non-nitrogenous to every one part of nitrogenous, while in animal foods the nitrogenous compounds are present in larger amount.

DRINKING WATER

Drinking water or potable water is water pure enough to be consumed or used with low risk of immediate or long term harm. In most developed countries, the water supplied to households, commerce and industry is all of drinking water standard, even though only a very small proportion is actually consumed or used in food preparation. Typical uses include washing or landscape irrigation. Over large parts of the world, humans have inadequate access to potable water and use sources contaminated with disease vectors, pathogens or unacceptable levels of toxins or suspended solids. Such water is not wholesome, and drinking or using such water in food preparation leads to widespread acute and chronic illnesses and is a major cause of death and misery in many countries. Reduction of waterborne diseases is a major public health goal in developing countries.

Water has always been an important and life-sustaining drink to humans and is essential to the survival of all organisms. Excluding fat, water composes approximately 70% of the human body by mass. It is a crucial component of metabolic processes and serves as a solvent for many bodily solutes. The

United States Environmental Protection Agency in risk assessment calculations assumes that the average American adult ingests 2.0 litres per day. Drinking water of a variety of qualities is bottled. Bottled water is sold for public consumption throughout the world.

Requirements

Health authorities have historically suggested at least eight glasses, eight fl oz each (240 ml), of water per day (64 fl oz, or 1.89 litres), and the British Dietetic Association recommends 1.8 litres.. This common misconception is not supported by scientific research. Various reviews of all the scientific literature on the topic performed in 2002 and 2008 could not find any solid scientific evidence that recommended drinking eight glasses of water per day. In the US, the reference daily intake (RDI) for water is 3.7 litres per day (l/day) for human males older than 18, and 2.7 l/day for human females older than 18 including water contained in food, beverages, and drinking water. The amount of water varies with the individual, as it depends on the condition of the subject, the amount of physical exercise, and on the environmental temperature and humidity. An individual's thirst provides a better guide for how much water they require rather than a specific, fixed quantity.

In terms of mineral nutrients intake, it is unclear what the drinking water contribution is. However, inorganic minerals generally enter surface water and ground water via storm water runoff or through the Earth's crust. Treatment processes also lead to the presence of some minerals. Examples include calcium, zinc, manganese, phosphate, fluoride and sodium compounds. Water generated from the biochemical metabolism of nutrients provides a significant proportion of the daily water requirements for some arthropods and desert animals, but provides only a small fraction of a human's necessary intake. There are a variety of trace elements present in virtually all potable water, some of which play a role in metabolism. For example sodium, potassium and chloride are common chemicals found in small quantities in most waters, and these elements play a role (not necessarily major) in body metabolism. Other elements such as fluoride, while beneficial in low concentrations, can cause dental problems and other issues when present at high levels. Water is essential for the growth and maintenance of our bodies, as it is involved in a number of biological processes. Profuse sweating can increase the need for electrolyte (salt) replacement. Water intoxication (which results in hyponatremia), the process of consuming too much water too quickly, can be fatal.

Access

Although covering some 70% of the Earth's surface, most water is saline. Freshwater is available in almost all populated areas of the earth, although it may be expensive and the supply may not always be sustainable. Sources where water may be obtained include:

- ground sources such as groundwater, hyporheic zones and aquifers.
- precipitation which includes rain, hail, snow, fog, etc.
- surface water such as rivers, streams, glaciers
- biological sources such as plants.
- the sea through desalination
- Water supply network.

Spring water, a natural resource from which much bottled water comes, generally contains minerals. Tap water, delivered by domestic water systems in developed nations, refers to water piped to homes through a tap. All of these forms of water are commonly drunk, often purified through filtration.

The most efficient way to transport and deliver potable water is through pipes. Plumbing can require significant capital investment. Some systems suffer high operating costs. The cost to replace the deteriorating water and sanitation infrastructure of industrialized countries may be as high as $200 billion a year. Leakage of untreated and treated water from pipes reduces access to water. Leakage rates of 50% are not uncommon in urban systems.

Because of the high initial investments, many less wealthy nations cannot afford to develop or sustain appropriate infrastructure, and as a consequence people in these areas may spend a correspondingly higher fraction of their income on water. 2003 statistics from El Salvador, for example, indicate that the poorest 20% of households spend more than 10% of their total income on water. In the United Kingdom authorities define spending of more than 3% of one's income on water as a hardship.

The Millennium Development Goal of halving the proportion of people without access to safe drinking water between 1990 and 2015 will probably be reached. Some countries though still face enormous challenges.

Rural communities are the furthest from meeting the 2015 MDGs drinking water target. Globally only 27% of the rural population has water piped directly to their home and 24% rely on unimproved sources. Of the 884 million people without access to an improved water source, 746 million people (84%) live in rural areas. Sub-Saharan Africa has made the least progress in improved water sources since 1990, improving only 9% to 2006. In contrast, the Eastern Asian region saw a dramatic drop from 45% to 9% reliance on unimproved water in the same time period.

In the U.S, the typical nonconserving single family home uses 69.3 gallons of water per capita per day. In some parts of the country there are water supplies that are dangerously low due to drought, particularly in the West and the South East region of the U.S.

Improving Availability

One of the Millennium Development Goals (MDGs) set by the UN includes environmental sustainability. In 2004, only forty-two percent of people in rural areas had access to clean water.

Solar water disinfection is a low-cost method of purifying water that can often be implemented with locally available materials. Unlike methods that rely on firewood, it has low impact on the environment.

One organisation working to improve the availability of safe drinking water in some the world's poorest countries is WaterAid International. Operating in 26 countries, WaterAid is working to make lasting improvements to peoples' quality of life by providing long-term sustainable access to clean water in countries such as Nepal, Tanzania, Ghana and India. It also works to educate people about sanitation and hygiene.

The Global Framework for Action (GF4A) is an organization that brings together stakeholders, national governments, donors and NGOs (such as Water aid) to define manageable targets and deadlines. 23 Countries are off-track to meet the MDG goals for improved water availability.

Well Contamination

Some efforts at increasing the availability of safe drinking water have been disastrous. When the 1980s was declared the "International Decade of Water" by the United Nations, the assumption was made that groundwater is inherently safer than water from rivers, ponds, and canals.

While instances of cholera, typhoid and diarrhea were reduced, other problems emerged. In India for example, 60 million people are estimated to have been poisoned by well water contaminated by excessive fluoride, which is dissolved from the granite rocks.

The effects are particularly evident in the bone deformations of children. Similar or larger problems are anticipated in other countries including China, Uzbekistan, and Ethiopia. Although helpful for dental health in low dosage, fluoride in large amounts interferes with bone formation.

In a related problem, it is estimated that half of the Bangladesh's 12 million tube wells contain unacceptable levels of arsenic due to the wells not being dug deep enough (past 100 M).

The Bangladeshi government had spent less than $7 million of the 34 million allocated for solving the problem by the World Bank in 1998. Natural arsenic poisoning is a global threat, 140 million people affected in 70 countries on all continents. These examples illustrate the need to examine each location on a case by case basis and not assume what works in one area will work in another.

Diarrhea as a Major Health Effect among Children

Over 90% of deaths from diarrheal diseases in the developing world today occur in children under 5 years old. Malnutrition, especially protein-energy malnutrition, can decrease the children's resistance to infections, including water-related diarrheal diseases. From 2000-2003, 769,000 children under five years old in sub-Saharan Africa died each year from diarrheal diseases. As a result of only thirty-six percent of the population in the sub-Saharan region having access to proper means of sanitation, more than 2000 children's lives are lost every day. In South Asia, 683,000 children under five years old died each year from diarrheal disease from 2000-2003. During the same time period, in developed countries, 700 children under five years old died from diarrheal disease. Improved water supply reduces diarrhea morbidity by twenty-five percent and improvements in drinking water through proper storage in the home and chlorination reduces diarrhea episodes by thirty-nine percent.

WATER QUALITY AND CONTAMINANTS

Parameters for drinking water quality typically fall under two categories: chemical/physical and microbiological. Chemical/physical parameters include heavy metals, trace organic compounds, total suspended solids (TSS), and turbidity. Microbiological parameters include Coliform bacteria, *E. coli,* and specific pathogenic species of bacteria (such as cholera-causing *Vibrio cholerae*), viruses, and protozoan parasites.

Chemical parameters tend to pose more of a chronic health risk through buildup of heavy metals although some components like nitrates/nitrites and arsenic can have a more immediate impact. Physical parameters affect the aesthetics and taste of the drinking water and may complicate the removal of microbial pathogens.

Originally, fecal contamination was determined with the presence of coliform bacteria, a convenient marker for a class of harmful fecal pathogens. The presence of fecal coliforms (like *E. Coli*) serves as an indication of contamination by sewage. Additional contaminants include protozoan oocysts such as *Cryptosporidium sp., Giardia lamblia, Legionella,* and viruses (enteric). Microbial pathogenic parameters are typically of greatest concern because of their immediate health risk.

Throughout most of the world, the most common contamination of raw water sources is from human sewage and in particular human faecal pathogens and parasites. In 2006, waterborne diseases were estimated to cause 1.8 million deaths each year while about 1.1 billion people lacked proper drinking water. It is clear that people in the developing world need to have access to good quality water in sufficient quantity, water purification technology and availability and distribution systems for water. In many parts of the world the only sources of water are from small streams often directly contaminated by sewage.

Safety Indicators

Access to safe drinking water is indicated by proper sanitary sources. These improved drinking water sources include household connection, public standpipe, borehole condition, protected dug well, protected spring, and rain water collection. Sources that don't encourage improved drinking water to the same extent as previously mentioned include: unprotected well, unprotected spring, rivers or ponds, vender-provided water, bottled water (consequential of limitations in quantity, not quality of water), and tanker truck water.

Access to sanitary water comes hand in hand with access to improved sanitation facilities for excreta. These facilities include connection to public sewer, connection to septic system, pour-flush latrine, and ventilated improved pit latrine. Unimproved sanitation facilities are: public or shared latrine, open pit latrine, or bucket latrine.

WATER TREATMENT

Most water requires some type of treatment before use, even water from deep wells or springs. The extent of treatment depends on the source of the water. Appropriate technology options in water treatment include both community-scale and household-scale point-of-use (POU) designs. A few large urban areas such as Christchurch, New Zealand have access to sufficiently pure water of sufficient volume that no treatment of the raw water is required.

Over the past decade, an increasing number of field-based studies have been undertaken to determine the success of POU measures in reducing waterborne disease. The ability of POU options to reduce disease is a function of both their ability to remove microbial pathogens if properly applied and such social factors as ease of use and cultural appropriateness. Technologies may generate more (or less) health benefit than their lab-based microbial removal performance would suggest.

The current priority of the proponents of POU treatment is to reach large numbers of low-income households on a sustainable basis. Few POU measures have reached significant scale thus far, but efforts to promote and commercially distribute these products to the world's poor have only been under way for a few years. In emergency situations when conventional treatment systems have been compromised, water borne pathogens may be killed or inactivated by boiling but this requires abundant sources of fuel, and can be very onerous on consumers, especially where it is difficult to store boiled water in sterile conditions and is not a reliable way to kill some encysted parasites such as *Cryptosporidium* or the bacterium *Clostridium*. Other techniques, such as filtration, chemical disinfection, and exposure to ultraviolet radiation (including solar UV) have been demonstrated in an array of randomized control trials to significantly reduce levels of water-borne disease among users in low-income countries, but these suffer from the same problems as boiling methods.

Drinking Water Regulation

Guidelines for the assessment and improvement of service activities relating to drinking water have been published in the form of International standards for drinking water such as ISO 24510.

European Union

The EU sets legislation on water quality. *Directive 2000/60/EC of the European Parliament and of the Council of 23 October 2000 establishing a framework for Community action in the field of water policy*, known as the water framework directive, is the primary piece of legislation governing water. The Drinking water directive relates specifically to water intended for human consumption.

Each member state is responsible for establishing the required policing measures to ensure that the legislation is implemented. For example, in the UK the Water Quality Regulations prescribe maximum values for substances that affect wholesomeness and the Drinking Water Inspectorate polices the water companies.

United States of America

In the United States, the Environmental Protection Agency (EPA) sets standards for tap and public water systems under the Safe Drinking Water Act (SDWA). The Food and Drug Administration (FDA) regulates bottled water as a food product under the Federal Food, Drug, and Cosmetic Act (FFDCA). Bottled water is not necessarily more pure, or more tested, than public tap water.

However, there is evidence that the United States federal drinking water regulations do not ensure safe water, as some of the regulations have not been updated with more recent science. Dr. Peter W. Preuss, who became the head of the U.S. EPA's division analyzing environmental risks in 2004, has been "particularly concerned", and has faced controversy in studies which suggest that regulations against certain chemicals should be tightened.

Drinking Water Preferences of other Animals

The qualitative and quantitative aspects of drinking water requirements of domesticated animals are studied and described within the context of animal husbandry. However, relatively few studies have been focused on the drinking behavior of wild animals. A recent study has shown that feral pigeons do not discriminate drinking water according to its content of metabolic wastes, such as uric acid or urea (mimicking faeces-or urine-pollution by birds or mammals respectively).

Water-soluble vitamins

Vitamin C or ascorbic acid

Absorption: the intestinal absorption of vitamin C is saturable, which means that it is useless to ingest more than 150 mg at a time. The excessive amounts that are sometimes ingested by vitamin C enthusiasts are mostly excreted and can result in diarrhoea and intestinal discomfort.

Metabolism, storage and excretion: the average half-life of ascorbic acid in the human body is 16 to 20 days. This time is inversely proportional to the intake and, thus, the catabolic rate is proportional to it. But the catabolism does not stop at the intake. Inevitable daily losses of several milligrams because of the irreversible degradation of ascorbic acid are observed. Hence the need to compensate for these losses to avoid fatal defi ciencies. There is no storage system specifi c to vitamin C. However, given an adequate intake of vitamin C, the cells of the organism contain enough of it to cope with a sudden drop in intake, and for clinical signs of deficiency to appear only two or three months later.

Function: the function of vitamin C is mainly related to its reversible role as chemical reducer. It enables the reduction of metallic ions necessary to certain reactions, and plays a major role in anti-oxidizing, and in the elimination of free radicals. It contributes to the formation of collagen (conjunctive support tissue), in addition to the synthesis of carnitine (transport for fatty acids) and neuro-transmitters. It also contributes to the metabolism of hormones, medicines and drugs, and the degradation of cholesterol. The vitamin promotes the absorption of iron, and mitigates the effects of chelating agents such as the phytates and tannins that form complexes that are insoluble with iron, thus preventing its absorption. The reduction properties of ascorbic acid could improve the stability of folic acid and vitamin E.

Deficiency: vitamin C deficiency results in a series of symptoms characteristic of scurvy. Scurvy is deadly if it is not treated rapidly with vitamin C.

Clinical, prophylactic and therapeutic considerations: apart from its anti-scurvy activity, vitamin C is thought to play an important role in the immune system, defence against oxidation, protection against cancer and the prevention of cardio-vascular disease.

Requirements: for adults, a daily dose of 10 mg of ascorbic acid treats scurvy, without allowing the subject to build vitamin reserves. The cost of pregnancy is estimated at 3 to 4 mg/day, whereas breastfeeding costs 15 to 45 mg/day on the basis of an average daily milk production of 750 ml. The needs of an infant fed on artifi cial milk are met with 7 mg of ascorbic acid per day. As of the age of 6 months, needs gradually approach the adult's. Quantities quoted here are adequate to treat scurvy; it is however unclear whether they suffi ce in relation to the many functions of vitamin C which, when reduced, do not result in immediate clinical evidence.

Pharmacological intake: 100 to 200 mg doses are recommended to address iron deficiency in exceptional need situations such as pregnancy, breastfeeding or antibiotic treatment. In case of injury, burns or surgical operation, vitamin C requirements could be in the vicinity of 1,000 to 2,000 mg to accelerate reparation and recovery.

Toxicity: when ingested in reasonable quantities (i.e. no more than 1 g/day), vitamin C has no toxic effect. The absorption of greater doses could/may entail negative effects, but available data remains unclear.

Thiamine or Vitamin $B_{1:}$ Source: thiamine is found in all animal and vegetable tissues – in other words, in all natural foods.

However, thiamine is not distributed equally in the tissues, and some refi nement processes can signifi cantly reduce its concentration in some foods.

The most abundant sources of thiamine are cereal sprouts, oilseeds and legumes, in addition to beer and pork. Fresh foods (greens, fruits, tubers, meat, milk and milk products except butter) contain enough thiamine to preclude deficiency if eaten regularly in suffi cient quantity.

Absorption: thiamine is absorbed actively when its concentration is inferior to 2 micromoles (ìmol) per litre, and passively when it is greater than 2 ìmol/l.

Metabolism, storage and excretion: the average half-life of thiamine in the organism is 9 to 18 days. If its concentration in the organism is excessive, it is not catabolized, but simply excreted as such. The approximate total quantity for the reference adult is 30 mg; this shows that thiamine is poorly stored in the organism, and that regular intake is indispensable.

Function: thiamine is a coenzyme that is indispensable to the energy metabolism, particularly for the junction between fermentation and breathing during the oxidation of glucides, and to the metabolism of amino acids whose carbon skeleton contains a branch. It also plays a role in nerve transmission, in terms of sodium and potassium transport.

Deficiency: thiamine deficiency causes a number of symptoms typical of beriberi. Beriberi is fatal if it is not treated rapidly with thiamine.

Requirements: thiamine requirements are mainly related to energy metabolism, i.e. energy use. They are therefore expressed in mg/1,000 kcal (4,180 kJ). The minimum required intake is of the order of 0.25 mg/1,000 kcal (4,180 kJ).

Pharmacological intake: thiamine-dependent patients only respond to pharmacological thiamine doses. This concerns only a few rare cases that respond to high thiamine intake, such as some forms of megaloblastic anaemia, lactic acidosis and branched chain ketoaciduria (or maple syrup urine disease, MSUD).

Toxicity: the organism is extremely effi cient in excreting thiamine excesses through urine; as a result, there is no evidence of thiamine toxicity provided it is administered orally.

Ribofl avin or Vitamin $B_{2:}$ Source: ribofl avin is found in most foods. Beer yeast, liver, eggs, milk, cheese and meat are its best sources, followed by green vegetables, and then fruits, legumes and, fi nally, cereals.

Absorption: ribofl avin is absorbed through an active saturable system. Absorption is facilitated by bile salts. The oral administration of ribofl avin as a medicine must coincide with meals, because food maintains it longer on absorption sites.

Metabolism, storage and excretion: in the tissues, ribofl avin is mostly found in the form of fl avine adenine dinucleotide (FAD), and most of it is free, and the rest is related to enzymes as a coenzyme. The liver is its main storage organ, and accommodates approximately one-third of the overall ribofl avin in the organism. As a coenzyme, FAD is quite stable, and its renewal is directly connected to the renewal of the proteins to which it is bound. When proteins are saturated, the free form is degraded into free ribofl avin, which is excreted mainly in the urine. Excretion is of the order of 200 ìg per 24 hours in reference adults, and falls by approximately 75% in case of deficiency.

Function: ribofl avin is a coenzyme that contributes to the oxidation-reduction of many metabolic pathways, particularly in energy production (ATP) in the respiratory chain.

Deficiency: ribofl avin is found in all major food categories, but in insuffi cient quantity in the essentially vegetarian diets of the poor in vast regions in the world, nor in traditional humanitarian relief food rations. As a result, moderate ribofl avin deficiency is common, combined with other more or less serious defi ciencies typical of nutritionally poor diets. The usual clinical signs include the oro-oculo-genital syndrome, together with angular stomatitis (sores at the corners of the mouth), glossitis (tongue infl ammation), cheilitis (swelling and cracking of the lips), a magenta colouring of the tongue, photophobia, and seborrhoeic dermatitis, particularly of the scrotum and the vulva. Of all these signs, only the magenta colouring of the tongue is an absolute indication of ribofl avin deficiency. Spotting it however requires a suffi ciently expert eye to distinguish it from other, close, colourings that indicate other defi ciencies. All the other clinical signs of ribofl avin deficiency can also result from other group B vitamin defi ciencies. Strangely enough, ribofl avin deficiency is not deadly, in spite of its fundamental role in the energy metabolism. This is probably because ribofl avin, once it is related to oxidation-reduction enzymes, is extremely stable, that it is found in all major food categories, and that a minimum amount is always present; in extreme food shortage situations, other defi ciencies are deadly sooner. The prescribed dose to deal with deficiency is 5 mg oral three times daily.

Requirements: as is the case for thiamine, the minimum ribofl avin intake is expressed in mg/1,000 kcal (4,180 kJ). Unlike thiamine, however, needs do not increase signifi cantly with energy use. The resulting minimum requirements are 0.5 mg/1,000 kcal (4,180 kJ)/day for infants. Additional needs related to pregnancy are 0.3 mg/day, and 0.5 mg/day for breastfeeding.

Pharmacological intake: because ribofl avin deficiency usually does not arise alone, it should be administered together with thiamine and, if possible, with all the other vitamins in therapeutic doses. A ribofl avin dose of 5 mg/ day is suffi cient, as research has shown no additional benefi ts in higher doses.

Toxicity: ribofl avin is non-toxic, and excessive intake is rapidly excreted in the urine.

Nicotinic acid and Nicotine-Amide (Nicotinamide) or Niacin or Vitamin B_3 *or PP:* The vitamin activities of nicotinic acid and its amide, nicotine-amide, are identical; they share the generic name of niacin.

Source: niacin is found in practically al l foods, if only in limited amounts. The highest concentrations are provided by meat (especially offal), fi sh, beer yeast, some legumes and some cereals. In the latter, niacin is blocked and becomes available only through alkaline treatment, such as the treatment of maize with lime water as practised by American Indians. Niacin has a precursor, or provitamin, from which it can be synthesized; this precursor is the essential amino acid tryptophan. Niacin synthesis from tryptophan requires ribofl avin and vitamin B_6. Tryptophan can contribute signifi cantly to niacin intake, provided it exceeds both the protein requirements and the homeostasis of amino acids: 60 mg of tryptophan in the food are generally considered to supply 1 mg of niacin. This fi gure nevertheless calls for caution: the conversion of tryptophan into niacin probably depends just as much on the content of one as of the other in the diet. Either way, food composition tables often refer to niacin equivalent, plus one-sixtieth of the tryptophan content of the food under consideration. This is a good refl ection of reality for foods that are poor in niacin but rich in tryptophan, such as milk and eggs that are indeed effective in addressing defi ciencies. Niacin in its free form is rare in nature, where it is essentially found as nicotinamide-adenine-dinucleotide (NAD) and nicotinamide-adenine-dinucleotide-phosphate (NADP). NAD and NADP are the two complete coenzymes whose active chemical group is nicotinamide.

Absorption: NAD and NADP are hydrolysed by enzymes of the intestinal mucous membrane, resulting in the release of the coenzyme nicotinamide. The intestinal fl ora partially converts nicotinamide into nicotinic acid, and both are then absorbed without it being known exactly how. The absorption is very effi cient, even in large quantities (approximately 3 g), with or without concomitant food intake. The tissues then rapidly extract niacin from the plasma.

Metabolism, storage and excretion: after the nicotinamide and nicotinic acid have been absorbed into the cells, NAD and NADP coenzymes form. Strictly speaking, there is no niacin storage in the organism. Excessive intake is rapidly excreted through the urine, where niacin is secreted as such.

Function: nicotinamide acts as an oxidant or a reducer by the exchange of a hydrogen atom with its substrate. More than 200 enzymes use either NAD or NADP as coenzyme. NAD is more frequently involved in catabolic reactions, using fuel molecules for the production of energy, whereas NADP is used rather in synthesis processes, such as the synthesis of steroids and fatty acids.

The conversion from NAD into NADP (and inversely) is easy, and the two molecules can exchange their oxidation states. Under hormonal control, this allows a balance to be maintained between synthetic reactions that consume energy and catabolic reactions that produce it.

Generally, niacin's oxidation-reduction function involves it in the metabolism (synthesis and degradation) of fatty acids, glucides, and amino acids.

Deficiency: niacin deficiency results in a number of symptoms typical of pellagra, which is deadly if it is not treated rapidly.

Clinical, prophylactic and therapeutic considerations: nicotinic acid prevents cardio-vascular disease by reducing plasma cholesterol levels, by also reducing low-density lipoproteins, and by increasing high-density lipoproteins. Niacin also plays an important role in the treatment of dementia, depression and schizophrenia, three illnesses that modify the niacin or tryptophan metabolism, resulting in increased requirements for this vitamin. Niacin is of course indispensable in the treatment of pellagra.

Pharmacological intake: a pharmacological niacin intake (3 g/day) is prescribed in the treatment of schizophrenia and of high plasma cholesterol levels.

Toxicity: niacin is not really toxic insofar as very high doses (3–6 g/day) only cause passing side effects. These can, however, be quite unpleasant: vasodilatation and hot fl ushes, nausea, vomiting, headache, and skin irritation.

Pantothenic acid or Vitamin $B_{5:}$ Source: pantothenic acid is found in all natural foods. It is particularly abundant in beer yeast, egg yolk, offal (especially liver) and is found in more moderate quantities in cereals and legumes. It is one of the constituents of coenzyme A, into which it is usually integrated. In food, it is found in both its free and integrated forms.

Absorption: the absorption of pantothenic acid appears to be passive, mainly in the small intestine, but possibly already in the stomach. The integrated form of coenzyme A must fi rst be released by enzyme action. The absorption of the vitamin is approximately 70%, the rest is excreted in the faeces.

Metabolism, storage and excretion: coenzyme A is synthesized in the cell from pantothenic acid. Following hydrolysis reactions, the latter is then released in free form and excreted as such in the urine; this means that pantothenic acid does not undergo any transformation in the organism. Pantothenic acid is not stored, excess intake is immediately excreted in the urine.

Function: coenzyme A is an indispensable vehicle of two-carbon units, both in the degradation of some amino acids, sugars, and fatty acids, and in synthesis. It is probably the most central and versatile of all coenzymes in the intermediary metabolism, while remaining just as indispensable to many more specifi c reactions, such as the synthesis of acetyl choline and porphyrines. Coenzyme A is also essential in the modifi cation of cellular proteins in terms of structure, activity, and location.

Deficiency: human pantothenic acid deficiency is rare, although moderate forms can evolve in the case of diets made up essentially of transformed foods from the food industry. Atypical symptoms include weariness, headache, and indisposition. The burning feet syndrome only develops in experimental deficiency conditions, or in grave forms of marasmus; its symptoms include the feeling of burning skin on the feet and ankles, that can extend up to the knee. The pain is sudden, spreads rapidly, and can be acute. Crises often occur at night and exhaust the patient by preventing him from sleeping. Cold water brings some relief. The burning feet syndrome is frequently found among malnourished prisoners whose diet is extremely monotonous and constantly unbalanced. It should be noted that the syndrome almost always appears in combination with the signs of other group B vitamin defi ciencies.

Pharmacological intake: there appears to be no indication for high doses of pantothenic acid. The burning feet syndrome is treated with an intramuscular dose of 10 mg/day. Administered orally, the vitamin also produces fast and excellent results, as an experience with severely malnourished prisoners has shown (fortuitously, a dose of 100 mg/day in a multivitamin preparation intended to treat beriberi was administered).

Toxicity: pantothenic acid is not toxic for humans.

Pyridoxine or Vitamin $B_{6:}$ Source: pyridoxine is found in most natural animal and vegetable foods. As is the case for most group B vitamins, it is abundant in baking yeast, liver and cereal sprouts. Avocado, banana, potato and some nuts and legumes also contain signifi cant concentrations.

Absorption: the absorption of pyridoxine is passive and non-saturable, and occurs in the small intestine.

Metabolism, storage and excretion: the liver provides the other tissues with active forms of pyridoxine (pyridoxal-5-phosphate), of which the major concentration is found in the striated muscle (80–90% of the overall pyridoxine contained in the organism). The vitamin is not specifi cally stored, and its turnover requires approximately 30 days. The pyridoxal-5-phosphate molecule is highly reactive, and excessive intake is irreversibly transformed into 4-pyridoxal acid, and eliminated in the urine.

Function: pyridoxine is primarily associated with practically all metabolic reactions of amino acids. It also contributes to gluconeogenesis, in the formation of niacin from tryptophan, in the synthesis of neuro-transmitters, in the regulation of steroid hormones, and in the compatibility between oxygen and haemoglobin. Its role in the lipid metabolism remains unclear and, generally, the function of pyridoxine is far from being comprehensively elucidated. In view of the vitamin's versatility, it is probably crucial for intake to be suffi cient and regular to ensure these multiple roles (both known and unknown), while avoiding that one function take precedence over another.

Deficiency: human pyridoxine deficiency is rare, because the vitamin is found in suffi cient amounts in all the major food categories. It is, however, associated with severe malnutrition or resulting from the administration of medicines that are incompatible with the vitamin, such as isoniazid, penicillamine, and oestrogen. Several illnesses related to an innate metabolic error cause a greater dependency on the vitamin.

Clinical signs among infants include convulsive crises, and anomalies in the electro-encephalogram; these are reversible through the administration of pyridoxine. Clinical signs among adults are those also associated with ribofl avin deficiency: stomatitis, cheilitis, glossitis (less typical tongue colouring), in addition to irritability, depression and confusion. The treatment consists of a 10 mg/day dose, administered orally.

Requirements: in view of the versatility of pyridoxine, its needs are diffi cult to defi ne in terms of a dose that would eliminate clinical deficiency signs. These signs can indicate deficiency with respect to specifi c functions, whereas others will prove insuffi cient at sub-clinical level. In this respect, specifi c data is lacking, and it is therefore necessary to limit action to recommended intakes.

Pharmacological intake: pharmacological pyridoxine doses could be effective against nausea and vomiting during pregnancy (50 mg/day), the withdrawal symptoms in the chronic alcoholic (100 mg/day), and the treatment of illnesses resulting from radiation (25–200 mg/day). Supplements are administered simultaneously to the treatment of tuberculosis by isoniazid, and Wilson's Disease by penicillamine (10 mg/day, orally).

Toxicity: pyridoxine is considered to be hardly toxic for man, in which no side effects are noted with pharmacological doses. The "low" toxicity aspect comes from the fact that disorders have been noted following massive intake. Once again, the consumer is responsible for his excesses.

Biotin or Vitamin B_8 or H or H_1: Source: biotin is synthesized by various micro-organisms, seaweed and plants that serve as food for species that cannot produce it. The richest food sources are beer yeast, liver, soy, and egg yolk. It should be noted that the human intestinal fl ora produces biotin, but it remains unclear to what extent this production is adequate to cover human needs and avert dependency on food sources.

Absorption: the absorption of biotin probably requires a specifi c transfer system.

Metabolism, storage and excretion: the biotin metabolism among humans is not well known, but its levels in plasma and in red blood cells tends to remain constant.

Function: biotin is crucial in the intermediary energy metabolism, involved both in the fuel synthesis (glucides/lipids) and its utilization to produce ATP.

Deficiency: human biotin deficiency is rare. Where it arises, it is usually related to an innate metabolic problem, massive raw egg white consumption, or the scientifi c experimentation of deficiency. Deficiency in these cases manifests itself mainly through seborrhoeic dermatitis. It is nevertheless possible that it may be involved in sudden infant death syndrome (SIDS, or cot death) in relation to artifi cially-fed infants. In any case, biotin deficiency is never a public health concern.

Requirements: the exact human biotin requirement is unknown. The recommended intake must therefore be deduced from the different known intakes preventing signs of deficiency or pathology in specifi c conditions.

Pharmacological intake: doses ranging from 10 to 100 ìg/day have been administered to treat specifi c biotin defi ciencies.

Toxicity: only experimental laboratory conditions achieve toxic biotin levels for animals; the pharmacological doses usual in parenteral (i.e. intravenous) human nutrition are well tolerated.

Folic acid (pteroylglutamic acid) or Vitamin $B_{9:}$ Source: folic acid is found in most natural foods; concentrated meat extracts and concentrated baking yeast are excellent sources, while offal, spinach, broccoli, asparagus, beetroot, cabbage, peanuts and avocado provide good sources. Wholemeal bread, eggs, rice, bananas and oranges provide moderate concentrations, and meat in general, grapes, apples, potatoes, and milk yield low concentrations. Losses due to cooking can be enormous, and the indications above therefore apply to the raw produce only.

Absorption: in most cases, the vitamin's natural form is tetrahydrofolic acid, which can be associated with up to six glutamate molecules. Absorption occurs mainly in the duodenum and the jejunum, following hydrolysis into monoglutamate. Absorption is active, stimulated by glucose, and saturable. However, high concentrations tend to result in passive absorption, thereby suggesting the existence of another transport system.

Metabolism, storage and excretion: the vitamin is found throughout the organism, with an overall amount of 5–10 mg, half of which is located in the liver. The folic acid metabolism is controlled by enzymes that regulate the balance between the monoglutamate (the usual form of transport) and polyglutamate forms (the active form of intra-cell storage); these proteins appear to manage a form of vitamin saving. In fact, the daily turnover rate of folic acid in the organism is less than 1%. High intakes of folic acid result in most of it remaining intact in the urine, together with 5-methyl-folic acid and 10-formyl-folic acid. However, experiments with radioactively marked vitamins show that, following major elimination of the intact form through the urine, the rest is recovered much more slowly in broken-down form. This suggests that excess folic acid is excreted rapidly, while the functional portion is fi nally broken down. Besides, some 100 ìg of active folic acid are excreted daily in the gall, most of which is absorbed again in the bowel, and the remainder is excreted in the faeces.

Function: folic acid is the acquiring and donor coenzyme of one carbon unit in the metabolism of amino acids and nucleotides, within a group of interdependent reactions where the vitamin's metabolism is cyclical, dependent on vitamin B_{12}.

Deficiency: megaloblastic anaemia is the best known manifestation of folate deficiency, and is common among the poor in tropical climates, particularly pregnant women, whose folic acid requirements are much higher. Deficiency can result from cooking habits that are deleterious to the vitamin, or from a reduced absorption due to gastro-intestinal disease or to alcohol consumption. However, it can also be associated with vitamin C, niacin and vitamin B_{12} defi ciencies. As always, real deficiency comes before – in this case, several months before – the clinical signs of anaemia, without obvious actual scope. It is nevertheless known that irritability, memory loss, hostility, and paranoid behaviour are also signs of folic acid deficiency.

Requirements: the folic acid provided by food is only available at approximately 50%: it is therefore necessary to specify whether the needs are expressed in folic acid from food, or synthetic folic acid. This Manual discusses folic acid from food. Adults require approximately 100 ìg per day to maintain a normal level of folic acid; pregnant women require approximately 675 ìg/day, and breastfeeding women between 200 and 300 ìg. Infants require approximately 10 ìg/kg/day.

Toxicity: folic acid is non-toxic for humans.

Cobalamin or Vitamin $B_{12:}$ Source: vitamin B_{12} is found in food as different forms of cobalamin. The long-held view that cobalamin is absent from vegetables is questioned today: some legumes may contain small amounts, although it is believed that these plants host micro-organisms that can synthesize them. The greatest concentrations of vitamin are found in offal (kidneys, liver and brain), less in egg yolk, oysters, crab and salmon. Other animal tissues and fi sh contain slighter concentrations. It should also be noted that the animal intestinal fl ora produces the vitamin, as this is the only means by which herbivores can obtain it. Human intestinal cobalamin production is insuffi cient, and diet must therefore supplement it.

Absorption: cobalamin is absorbed in the ileum, at fi rst actively through the connection of cobalamin to the intrinsic factor, which is a glycoprotein molecule synthesized by the parietal cells of the gastric mucous membrane. This process is indispensable to the absorption of the vitamin at the levels usually found in food. There also exists a dissemination process that does not involve the intrinsic factor, whereby the vitamin is administered orally, as a medicine, in doses that are higher than those found in food. Metabolism, storage and excretion: the organism does not break down cobalamin, which is found intact in the faeces and urine. Excess intake is stored in the liver, which contains more than 50% of the 2 to 5 mg of vitamin usually found in the organism. This amounts to quite a high storage in comparison to other vitamins, because its excretion is low: slightly more than 1 ìg/day through urine, whereas approximately 70% of the vitamin reaching the digestive tube through gall is absorbed again. The resulting daily excretion is estimated at between 2 and 5 ìg: this is equivalent to an autonomy of almost three years, not to mention the synthesis through the intestinal fl ora.

Function: vitamin B_{12} is essential in the regeneration of 5-methyl-folic acid into folic acid, combined with the formation of methionine amino acid. Without this regeneration, the metabolic journey of 5-methyl-folic acid inevitably ends in excretion through urine. This causes a deficiency in another intermediary of the metabolic cycle of folic acid; this intermediary contributes to the synthesis reaction of a nucleic acid itself indispensable for the synthesis of desoxyribonucleic acid (DNA). This metabolic trap of folic acid, in the case of vitamin B_{12} deficiency, explains why anaemia related to this deficiency is impossible to distinguish from anaemia related to folate deficiency. Vitamin B_{12} furthermore contributes to the transport and storage of folic acid in the cells. It plays a signifi cant, albeit still unclear, role in myelin synthesis; it also contributes to the metabolism of glucides, lipids and proteins as a reducing agent.

Deficiency: cobalamin deficiency is caused either by insuffi cient absorption due most frequently to the incapacity to produce the intrinsic factor (but also, sometimes, to other ailments of the digestive tube), or by insuffi cient vitamin intake.

Incapacity to produce the intrinsic factor causes pernicious anaemia, which is deadly if it is not treated; it also presents impairments typical of nutritional anaemia, in addition to neuropathy (most frequently, fi rst in the peripheral nerves, followed by the spinal cord and the brain). The clinical signs of nervous impairment include tingling and a feeling of cold and numbness in the extremities, followed by muscular weakness and lack of movement coordination as the spinal cord is affected. Mental symptoms can also appear: intellectual slowness, memory lapses, confusion, depression and paranoid psychoses. The diagnosis of nutritional anaemia should include the verifi cation that it is a cobalamin and/or folate deficiency: on the one hand, the two vitamins are not interchangeable in the treatment of nutritional anaemia and, on the other, the administration of folic acid can improve pernicious anaemia momentarily, while the neurological damage increases, sometimes irreversibly. The treatment of pernicious anaemia leads to total remission, provided there is no irreversible neurological damage yet; it consists of a monthly subcutaneous or intramuscular injection of 100 ìg of vitamin B_{12}, for life.

Insuffi cient vitamin B_{12} intake is rare and only concerns those too poor to secure animal products, or those confi ned to a strict vegetarian diet by their religion or their personal principles. Deficiency develops slowly, since the intestinal fl ora produces the vitamin, but in amounts that are insuffi cient to satisfy the need. When the dietary insuffi ciency results from poverty, anaemia and neuropathy develop equally, because the diet is poor in cobalamin and folic acid alike. When the insuffi ciency is caused by cultural restrictions, anaemia is slight, even absent, while neuropathy can be severe, because a deliberately vegetarian diet is usually rich in folic acid, and this compensates the metabolic trap effect affecting 5-methyl-folic acid. This phenomenon also appears in pernicious anaemia in richer wealth groups that develop neuropathy primarily, and in poorer strata that develop both neuropathy and anaemia: the diet of the former tends to be richer in folic acid than the latter's. The treatment of deficiency caused by insuffi cient intake consists of the introduction of animal produces in the diet, or the administration of vitamin preparations that supply orally 1 ìg of vitamin B_{12} per day. Another solution is the treatment of pernicious anaemia.

Requirements: a daily 1 ìg dose has been observed to contain anaemia and neuropathy entirely among adult patients suffering from anaemia due to dietary vitamin B_{12} deficiency. The need is proportionately greater among growing children, particularly among infants.

Retinol, axerophtol or Vitamin A: Source: vitamin A is found in food, mainly bound to fatty acids, the most common of which is palmitic acid. In order of decreasing concentration, it is found in sea fi sh and sea mammal liver oils (in large quantity), in the liver of animals for slaughter and poultry, butter, cheese, wheat germ, eggs and oily fi sh, fi nally in offal other than liver, milk, lean fi sh, red and white meat. Retinol (vitamin A alcohol) has some precursors or provitamins, mainly â-carotene, which is the main representative of the large carotenoid family. â-carotene is a retinol dimer that releases two retinol molecules per central splitting of the â-carotene molecule. It appears that â-carotene can also break down into retinaldehyde through eccentric splitting, without in this case producing two retinol molecules, but one only. Retinaldehyde is easily reduced into retinol in the presence of NADH.

â-carotene is found in vegetables and coloured fruits, among others: spinach, parsley, the leaves of cassava, amaranth and lettuce, carrots, turnips, cabbage, tomatoes, mandarins/tangerines, melons, oranges, dates, papayas, mangoes; in butter, palm oil (the richest source), wheat sprouts and fl our, yellow maize, lentils and yellow sweet potatoes. According to the ingested amounts, the biological activity of the different provitamins is inferior to that observed for retinol, because food carotenoids are less well digested and absorbed than vitamin A; furthermore, carotenoids are not entirely transformed into retinol. They are then expressed as retinol equivalents (RE): 1 RE = 1 ìg of retinol, or 6 ìg of â-carotene, or 12 ìg of other carotenoids acting as provitamins. The 1 to 6 and 6 to 12 relationships are estimated, and are used for want of a better reference. Formerly (but old habits die hard), vitamin A was also expressed in international units (IU): 1 IU amounts to the activity of 0.3 ìg of retinol, or to 0.55 ìg of retinol palmitate. Attention should nevertheless be paid to the fact that there are also IU of provitamin A, where 1 IU = 0.6 ìg of â-carotene. To clarify, retinol IU are labelled IU_a and â-carotene IU are labelled IU_c. The retinol to â-carotene ratio is 1 ìg to 6 ìg in terms of RE, and 1 ìg to 2 ìg in terms of IU. This is due to the fact that experiments that led to the defi nition of â-carotene IU were conducted in optimal conditions, neglecting many a factor that determines what ultimately remains in the form of usable retinol in the organism from a given quantity of ingested â-carotene.

Absorption: in the stomach, vitamin A and its provitamins combine with the other lipids to produce globules that pass into the small intestine. There, globules combine with bile salts to produce emulsions that permit the splitting of complex lipids into simple lipids owing to the action of digestive enzymes; these emulsions then form micelles of mixed lipids that enter into contact with the brush-shaped border of the epithelial cells of the intestinal mucous membrane. Vitamin A is transported through the intestinal wall through a specifi c system when it is in low concentrations, and by passive dissemination when its concentration is high, whereas provitamins disseminate passively. The absorption of vitamin A and its precursors is facilitated by the presence of proteins and lipids in the diet. The state of the intestinal mucous membrane is of critical importance. Bile salts are indispensable to the absorption of carotenoids, whereas vitamin A does not depend upon it (provided that it is properly soluble). Approximately 80% of the retinol contained in the diet is absorbed (this effi ciency only diminishes with very high doses of retinol), whereas only 50% of the provitamins is absorbed, and this effi ciency drops quickly with high doses.

Metabolism, storage and excretion: after its absorption, retinol is transported by chylomicrons to the liver in the form of retinol palmitate. Carotenoids are mostly broken down into two retinaldehyde molecules by central splitting, in the intestinal mucous membrane, and then transported to the liver via chylomicrons. However, a signifi cant amount of carotenoids can be incorporated as it is into chylomicrons, and then be transported and stored in other tissues than the liver (mainly adipose tissue). Retinol is either stored in the liver as retinol palmitate, or transported to other tissues by means of the retinol binding protein (RBP). Within the tissues, retinol is also bound to cellular retinol binding proteins (CRBP). Generally speaking, ingested vitamin A takes the following course: some 20% are directly excreted in the faeces in the following two days, 20% to 50% of the absorbed 80% are excreted in the faeces and urine during the week following ingestion, and the remainder is stored. Vitamin A reserves are considerable in adults: with a normal initial stock, they can live for one to three years without additional intake. Infants and children, on the other hand, only have minimal retinol reserves. Stored vitamin A is only metabolized very slowly, in two ways: by RBP-retinol degradation in the kidneys and peripheral epithelial tissues (the skin and the mucous membranes lining the internal body surfaces), or by retinol transformation in the liver and peripheral tissue, in combined or oxidized forms that are later excreted in the faeces or urine.

Function: vitamin A contributes to vision, cell differentiation, morphogenesis, in addition to complex physiological processes such as growth, reproduction, and the immune response.

Generally speaking, most tissues probably rely more or less on vitamin A, which explains the increasing recognition of its major signifi cance.

- Vision: vitamin A is one of the indispensable molecules for the transmission of information registered by the eye to the optical nerve.
- Cell differentiation: vitamin A deficiency is observed to cause cells that produce mucus to be replaced by cells that produce keratin. The introduction of vitamin A into culture cells rapidly causes various types of cell lines to differentiate. The effect of vitamin A, or rather its deficiency, is particularly obvious on the cells of the cornea, which keratinises and fi nally ulcerates.
- Morphogenesis: both the excess and the deficiency of vitamin A impair embryogenesis, but the mechanism is unknown.
- Growth: vitamin A is a growth factor, through its derivative, retinoic acid. Here again, the mechanism is unknown. One of the fi rst signs of deficiency is appetite loss, but it is not the real cause of growth problems related to vitamin A deficiency.
- Reproduction: sperm formation (spermatogenesis) and foetal growth are affected by retinol deficiency, probably owing to the role of vitamin A in cell differentiation.
- Immune response: vitamin A is also referred to as "anti-infection vitamin". Specifi c and non-specifi c defence mechanisms both depend on it. Its deficiency, combined with protein-energy malnutrition has major consequences on the organism's resistance to infection.

Deficiency: among others, vitamin A deficiency leads to blindness and high mortality risks from infections such as measles. It should be noted, however, that any imbalance generally has a negative effect on the functions of vitamin A. Protein deficiency reduces the â-carotene cleavage and RBP synthesis; on the other hand, protein-energy deficiency restricts growth, and thus postpones the appearance of vision disorders caused by vitamin A deficiency. To be absorbed, vitamin A and retinoids must be ingested together with lipids. Iron levels depend on vitamin A, whereas the utilization of retinol is related to zinc levels. The vitamin A metabolism depends to a certain extent on vitamin E. Moreover, vitamin A interacts with vitamins C, K, and D, calcium, copper, and iodine.

Chylomicrons: microscopic lipid particles common in the blood during fat digestion and assimilation, consisting of 98% lipids and 2% proteins.

Requirements: experimentation has defi ned minimum male adult vitamin A requirements to be approximately 400 ìg/day.

Pharmacological intake: retinoic acid and, to a lesser extent, carotenoids, are used at high dose in the treatment and prevention of a number of illnesses such as acne, skin cancer (but not melanoma), acute myeloid leukaemia (AML), and breast cancer. â-carotene intake is directly associated with low lung cancer incidence.

Toxicity: taken in high doses, vitamin A is toxic to the point of being lethal. Three different types of toxicity should be distinguished, as follows.

- Acute toxicity follows the ingestion of a single or several high doses. The threshold dose for risk of acute toxicity begins at 200 mg of retinol for adults, 100 mg for children above 5, 60 mg for small children, and 30 to 40 mg for infants. Its signs appear in the hours following absorption: nausea, vomiting, headache, dizziness, double vision, lack of coordination and muscular weakness. Fontanel swelling is possible in infants. Very high doses frequently cause vomiting, itching around the eyes, and fl aky skin, particularly around the mouth. These signs usually subside within a few days, once the ingestion of massive doses has been discontinued. By extrapolation, studies with apes (which seem to provide excellent models for vitamin A toxicity in humans) show that lethal doses would be in the vicinity of 500 mg for infants, and 11.8 g for adults (Shils, 1994).
- Chronic toxicity is quite common, because of the sometimes exaggerated popularity of vitamins, which are often imposed upon children with the best intentions. Daily intakes that can cause chronic toxicity are of the order of 10 times the recommended intake, that is, 3.75 mg for the infant, and 10 mg for adults. The most common signs of chronic intoxication include the following: skin dryness and colouring, headaches, anorexia and weight loss, alopecia (hair loss), articular stiffness and pain, and fatigue. It takes weeks, even months, for these signs to subside once normal vitamin intake has been resumed.
- Teratogenesis: vitamin A is highly teratogenic. In early pregnancy, a single dose superior to 200 mg or daily doses exceeding 30 to 90 mg for 1 week, or long-term daily intakes in the order of 7.5 mg of retinol can cause miscarriage or severe malformations of the foetus (Shils, 1994). Vitamin A supplements are therefore not recommended during pregnancy (particularly during the fi rst quarter) for healthy women whose regular diet includes fruits and greens.

Either way, intake should not exceed 3 mg of retinol per day. Carotenoids, on the other hand, are not toxic even in very high doses because their absorption is lower than retinol's, and is quickly saturated.

Cholecalciferol or Vitamin $D_{3:}$ Source: vitamin D_3 is not really a vitamin, because it is synthesized by man through solar ultraviolet radiation on his skin. Ultraviolet irradiation transforms provitamin D_3 (7-dehydrocholesterol) into an intermediary, which spontaneously converts into vitamin D_3 within one or two days simply because of the organism's heat. In adults, synthesis occurs mainly in the outer layer of the skin (epidermis), whereas in infants it also occurs in the deep inner layer of the skin (dermis), because their epidermis transfers more ultraviolet radiation to the dermis. Prolonged exposure to the sun does not cause hypervitaminosis, because both the provitamin and the vitamin are rapidly broken down into inactive biological compounds. The main source of vitamin D_3 comes from endogenous synthesis. However, the vitamin is found in high concentrations in cod liver oil, and is also found

Data for other population categories is not available; readers are thus referred to recommended intakes. Teratogenesis: a prenatal toxicity characterized by structural or functional defects in the developing embryo or foetus. in sea fi sh, egg yolk and dairy products. Following the rickets epidemics that affected the industrial cities of the northern United States and Europe mainly, vitamin D_3 or D_2 enrichment is common in milk, margarine and cereals (fortifi ed foods vary according to country). Vitamins D_3 and D_2 are analogous, and will hereafter simply be referred to as vitamin D. Vitamin D has two types of unit: micrograms (ìg) and international units (IU), 1 IU amounting to 0.025 ìg, and 1 ìg amounting to 40 IU.

Absorption: approximately 80% of the vitamin D supplied by the diet is absorbed, but absorption is highly sensitive to the state of the intestinal mucous membrane, and to its capacity to absorb fats. It is transported by chylomicrons to the liver. The endogenous vitamin travels to the dermal capillary network through a specifi c transport protein, and is then transported to the liver.

Metabolism, storage and excretion: the liver modifi es vitamin D into hydroxycholecalciferol, which is converted in turn by the kidney into dihydroxycholecalciferol, the physiologically active form of the vitamin. Vitamin D, or cholecalciferol, is stored in adipose tissue, and hydroxycholecalciferol is stored in the muscle. Blood holds the highest concentration of vitamin D and, unlike vitamin A, the liver does not store it. Little is known about the turnover and excretion process of vitamin D.

Function: vitamin D appears to function like a steroid hormone. Its most striking role is to maintain the calcium concentration within and around the cell by modulating the calcium and phosphorus metabolism in the intestine and the bones. Vitamin D does not contribute directly to ossifi cation, but promotes it by maintaining oversaturated calcium and phosphorus concentrations around the cell (i.e. extra-cellular). Apart from its action on minerals, vitamin D also contributes to cell multiplication and differentiation. It is effective in the treatment of psoriasis, a chronic, recurrent skin disorder.

Deficiency: vitamin D deficiency arises from poverty, poor housing, pollution, inadequate clothing, and an over-protection against solar radiation (for fear of skin cancer). It manifests itself in children through rickets, and in adults through the softening of bony tissue (osteomalacia). In the late 19th century, up to 75% of poor children in industrial cities were rachitic (Davidson, 1979). Narrow streets, high corbelled buildings, and domestic and industrial fumes all combined to reduce ultraviolet solar radiation signifi cantly. Moreover, the poor lacked the means to purchase foods rich in vitamin D foods such as dairy products and eggs and, in the cities, had little access to fresh air.

These causes for rickets have subsided, but have by no means disappeared today; sub-clinical forms of rickets still (or again) constitute a public health problem in Western countries. Rickets is also related to protection and dressing customs for children, and the confi nement of women and small children to backyards. In parts of Ethiopia, for instance, it is customary for infants be covered from head to foot when leaving the house, up to the age of approximately 6 months, so as to be protected from negative external infl uence that could make them sick. Rickets is therefore common among very small children, but recedes rapidly as soon as the child is left to move outside uncovered.

Vitamin D deficiency results in insuffi cient intestinal absorption, and insuffi cient re-absorption of calcium and phosphorus in the kidney. The serum protein levels of both minerals drops, resulting in the case of calcium in hyperparathyroidism (the excessive secretion of parathyroid hormone by one or several parathyroid glands). The parathyroid hormone and the remaining vitamin D then cause the demineralization and weakening of the bones. This leads to the malformations typical of rickets: bowlegs, knock-knees, a deviation of the spine, a beady appearance of the ribs (rachitic rosary), and narrowed chest and pelvis. These malformations are purely mechanical, because the bones lack the resilience to sustain normal strain. Furthermore, rachitic children are agitated, worried and irritable, are pale, perspire abnormally from the head, their muscles are soft and lack tonus, leading to a distension of the abdomen, and frequently have diarrhoea.

Hypocalcaemic tetany (spasms of the extremities caused by calcium deficiency in the blood) can also occur. Rickets itself is not deadly, but increases the risk of mortality owing to an increased liability to respiratory infection. Its prognosis is nevertheless fairly good, because bone deformations tend to correct themselves spontaneously during growth.

Osteomalacia is the manifestation of vitamin D deficiency in the adult. It affects mainly women of childbearing age whose calcium levels have been weakened by repeated pregnancies. Problems are aggravated if they have suffered from rickets in their childhood, if social customs require them to remain indoors, or if they come from poor backgrounds whose access to food is inadequate. The elderly are also vulnerable to this deficiency because of their often neglected diet, their physical and psychological limitations in going out, and their reduced intestinal absorption of calcium; osteomalacia is in this case frequently combined with osteoporosis. Osteomalacia causes muscular weakness, bone pains and pelvic deformations that result in serious obstetric complications in women of childbearing age. Deformations resulting from osteomalacia are practically irreversible, or only through orthopaedic surgery.

The treatment of rickets and osteomalacia consists of daily doses of 25 to 125 \ig of vitamin D for 1 month, in addition to 1 to 2 g of calcium lactate per day, in order to re-mineralize the bones and avert hypocalcaemic tetany. Measures must then be taken to promote exposure to the sun and improve the diet; this is usually extremely diffi cult, as it implies addressing strong cultural values and poverty. In Western countries, the decline in school and domestic education, and the alarming increase of poverty may well cause a fresh upsurge of rickets.

Requirements: the minimum daily requirement to prevent rickets and osteomalacia is 100 IU (2.5 \ig). This does not mean much, as the proportion of endogenous synthesis is unknown. Exposure to the sun and diet vary considerably, and a daily intake in the food is therefore recommended today. It should be 400 IU (10 \ig) for children, and 100 IU (2.5 \ig) for adults.

Toxicity: vitamin D quickly becomes toxic, that is, as of doses 5 times the recommended intake. The result is hypercalcaemia in children and nephrocalcinosis in adults. In children signs of intoxication include appetite loss, often combined with nausea and vomiting. The child loses weight, is irritable and becomes depressed. Intoxication can be fatal.

Tocopherols or Vitamin E: Introduction: vitamin E is the generic reference to 8 related molecules (isomers) whose biological activities differ; the four most common are oc-tocopherol, (3-tocopherol, ytocopherol and 5-tocopherol. Little is still known about vitamin E. Various signs of deficiency have been documented for animals, but human deficiency appeared to be non-existent - vitamin E has only recently proven to be a real vitamin, and thus essential, for man.

Source: octocopherol and y-tocopherol are the forms most commonly found in nature, mainly in wheat germ, sunfl ower, saffl ower, rapeseed, groundnut, corn and olive oils. Other animal and vegetable foods on average only contain approximately 5% of the vegetable oil content. Some oils, such as soy oil, contain considerably more y-tocopherol than oc-tocopherol. oc-tocopherol is the isomer that contributes most to vitamin activity. (3-tocopherol, y-tocopherol and 5-tocopherol respectively contain only 40 to 50%, 10 to 30% and approximately 1% of the activity of oc-tocopherol. This once again leads to the use of international units, with the following values:

oc-tocopherol: (3-tocopherol: y-tocopherol : 5-tocopherol:

1.49 IU/mg 0.60 IU/mg 0.15 - 0.45 IU/mg1 0.015 IU/mg

1 IU = 0.67 mg 1 IU = 1.67 mg IU = 6.67 - 2.22 mg 1 IU = 66.7 mg

It is diffi cult not only to establish food composition tables but also to utilize them for any given diet because of the variable activity of the different isomers, their instability to heat, oxygen and some food preparations, and the seasonal variations of their concentration in food. The real or effective vitamin E diet content generally has little to do with values contained in tables.

Absorption: the absorption of vitamin E occurs mainly in the middle area of the small intestine. Normal dietary intake leads to the absorption of approximately 50% of vitamin E; this proportion is inversely proportional to the ingested amount. The presence of lipids in the diet promotes absorption, which also depends on bile salts and pancreatic enzymes. At this stage, no further reliable knowledge exists as to vitamin E absorption. Metabolism, storage and excretion: what is known about the vitamin E metabolism largely comes from animal research, in particular rats. The vitamin concentration in tissue rises with intake; on the other hand, different tissues and organs accommodate very different vitamin E concentrations, without it being clear why. The vitamin accumulates especially in membrane-rich cell fractions. The organs that contain most vitamin E are adipose tissue, the liver and muscle, and these are considered to be storage organs. Little is known about the vitamin E metabolism: it appears to be very slow, and the excess vitamin is excreted as such in the faeces.

Function: current knowledge suggests that vitamin E functions as an antioxidant in cell membranes. In particular, it is thought to protect polyunsaturated fatty acids from free radicals, to which the former are highly vulnerable. Some signs of vitamin E deficiency in animals have disappeared completely following the administration of antioxidants.

Deficiency: the only certainty regarding human vitamin E deficiency is the increasing sensitivity of red blood cells to *in vitro* haemolysis, and their reduced *in vivo* lifespan when plasma vitamin E levels fall below 0.5 ml/dl.

For want of a better threshold, plasma vitamin E levels lower than 0.5 ml/dl are considered to indicate deficiency. Preterm (or premature) infants and most patients suffering from intestinal malabsorption (for whatever reason) fall in this category. Vitamin E deficiency was formerly suspected of causing specifi c pathologies among preterm infants, such as retinopathy of prematurity (ROP) and haemolytic anaemia. However, recent research appears to contradict these hypotheses. In patients suffering from malabsorption, the relation between low plasma vitamin E levels and specifi c disorders, such as Duchenne's myopathy, could not be demonstrated either. On the other hand, the lifespan of erythrocytes is reduced in these patients, and this can be corrected by the daily administration of 100 to 200 IU of vitamin E. This confi rms the physiological role of vitamin E, and its inclusion among essential nutrients. Recommended daily therapeutic doses in case of malabsorption are 25 to 50 IU for infants, 50 to 100 IU for children from 1 to 10 years, 100 IU for children and teenagers from 10 to 18 years, and 200 IU for teenagers and adults above 18.

Requirements: the daily minimum adult requirement to maintain plasma tocopherol levels above 0.5 ml/dl has been shown to be 2 mg of á-tocopherol (H" 3 IU).

Pharmacological intake: there is currently no concrete evidence that pharmacological doses of vitamin E provide any benefi t whatsoever.

Toxicity: no toxicity has been observed in cases of excessive daily vitamin E intake, such as 100 to 800 mg; this may simply be due to the fact that at such levels, most of the vitamin is probably not absorbed. Vitamin E is however assumed to be toxic and, because tocopherol accumulates in tissues according to intake, caution is called for especially in case of intravenous vitamin administration.

CARBOHYDRATES

Carbohydrates are a major source of energy for humans, providing approximately 45% to 80% of the total caloric intake in different income groups. Since they are a relatively inexpensive source of energy compared to fats and proteins, they form the bulk of the diet of humans throughout the world. They are mainly present in food in the form of sugars, starches, and fibres. A study of the various types of carbohydrates is necessary because the kind and pro-portion of different forms of carbohydrate present in food have a direct bearing on our health. Three groups of carbohydrates are of importance in our diet from the nutritional point of view, namely, sugars, starches, and fibres.

The sugar and starch that we consume is ultimately broken down to glucose in the digestive tract and absorbed into the blood circulation. In the human body, glucose is removed from blood by the tissue cells and used as a source of energy. Some glucose is converted to glycogen, also called animal starch, and stored in the musicle and liver as a reserve store of energy.

CLASSIFICATION OF CARBOHYDRATES

Carbohydrates which are of importance in the diet are classified on the basis of the number of sugar units present in them. They may also be classified as:

Available carbohydrates: Carbohydrates which can be digested in the human body and yield energy when they are oxidized in the body.

Unavailable carbohydrates: Carbohydrates which cannot be digested because the human body does not contain the enzymes necessary for their breakdown. Unavailable carbohydrates do not provide any energy to the body but are necessary as they perform some important functions in the body such as regular elimination of faecal waste.

DIGESTION, ABSORPTION, AND METABOLISM

Complex carbohydrates and sugars are too large to be absorbed through the intestinal wall. They need to be broken down into their constituent monosaccharides so that they can be absorbed. Only monosaccharides can be absorbed into the blood stream. The mechanical and chemical digestion of starch begins in the mouth.

Ptyalin or salivary amylase in saliva acts on cooked starch and partially breaks it down into dextrin. If ptyalin acts on starch long enough, i.e., if food is chewed well, some maltose may be formed. The stomach does not secrete any starch-splitting enzyme, hence no digestion takes place in the stomach. In the small intestine, pancreatic amylase acts on starch and dextrin breaking it down into maltose.

The intestinal wall secretes maltase, lactase, and sucrase which acts on maltose, lactose, and sucrose, reducing them to their respective monosaccharides. Some fructose may be converted to glucose. The monosaccharides are absorbed into the blood vessels lining the small intestine and carried to the liver for their further metabolism.

The absorbed monosaccharides, i.e., glucose, fructose, and galactose may be converted into glycogen and stored in the liver, or converted into glucose and released into the blood stream to be oxidized as a source of energy for various tissue cells.

In human metabolism, all sugars are converted into glucose. In the muscle cells, some glucose may be stored as glycogen. The fasting level of glucose in blood is maintained at 70-100 mg/100 ml blood. After consuming a meal rich in carbohydrates, it increases to 140-150 mg/100 ml blood. Glucose is taken up from blood by the body cells and oxidized as a source of energy.

In the cells, glucose is first oxidized to form pyruvic acid, and ultimately through the various metabolic cycles, energy is released in the form of adenosine triphosphate (ATP). The waste products of carbohydrate metabolism, i.e., carbon dioxide and water are released from the cell and excreted by the body. The energy released is used by the body for its various voluntary and involuntary processes and to maintains body temperature.

If the level exceeds 170 mg glucose/100 ml blood, it crosses the renal threshold and glucose is excreted in the urine. This condition is observed in diabetic patients. A fasting blood glucose level above 140 mg is called hyperglycaemia and below 70 mg is called hypoglycaemia.

The body tries to maintain the normal fasting level by remov-ing glucose from blood when the level is high and adding glucose to blood when the level falls, e.g., when a person is fasting. The hormone insulin, secreted by the cells of the islets of Langerhans in the pancreas, helps in regulating the blood glucose level. Insulin is required for glucose utilization by the cells and synthesis of glycogen from glucose. If insulin is deficient, glucose is not utilized by the cells and blood level of glucose increases. Insulin is the only hormone which lowers blood sugar levels, other hormones, such as the thyroid hormone, increase blood sugar levels.

Sources

Daily diet should provide up to 50-70% kcal of energy from carbohydrate, which means that the diet of an individual who needs 2,400 kcal should consume 60% of 2,400, i.e., 1,440 kcal or 360 g of carbohydrates/day. Carbohydrates are not only an economical source of energy but are also readily available and easy to store as they have a long shelf life.

All foods of plant origin contain carbohydrates in varying amounts. With the exception of milk, animal foods do not contain carbohydrate. Although milk is not consumed as a source of carbohydrate, some milk products, such as khoa and milk powder, contain significant amount of carbohydrate lactose.

The important sources of carbohydrates in the diets of Indians are cereals and millets, roots, tubers, pulses, sugar, and jaggery.

All sugars provide 4 kcal/g of energy. The carbohydrate and calorie content of a food can be reduced by using sugars which are sweeter than sucrose so that the quantity of sugar required will be less.

Functions

Carbohydrates have many functions in the human body.

1. The chief function of carbohydrate is to provide energy to the body so that it can carry out day-to-day work and maintain body temperature. All carbohy-drates except fibre provide 4 kcal/g of energy. It is the cheapest source of energy available.
2. Glucose is the only form of energy used by the central nervous system. When blood glucose levels fall, the brain does not receive energy and convulsions may occur.
3. Carbohydrates spare proteins from being broken down for energy and are used for bodybuilding and repair. In carbohydrate deficient diets, proteins meant for bodybuilding and repair are oxidized to meet the most important and first need of the body, i.e., energy.
4. They are required for complete oxidation of fat. In a deficiency, fats are broken down rapidly for energy and intermediate products such as ketones are formed in large amounts resulting in a condition called ketosis.
5. Carbohydrates can be converted into non-essential amino acids, provided a source of nitrogen is available.
6. The sugar lactose helps in the absorption of the minerals calcium and phosphorus.
7. Lactose helps certain bacteria to grow in the intestine. This bacterial flora is capable of synthesizing B-complex vitamins in the gut.
8. Dietary fibre plays an important role of increasing faecal mass by absorbing and holding water, stimulating peristalsis, and eliminating faecal waste.
9. Fibre also helps in lowering blood cholesterol levels by binding bile acids and cholesterol.

Deficiency

The daily diet should not contain less than 100 g of carbohydrate. Carbohydrate deficiency is uncommon in our country as diets are cereal based. A deficiency of carbohydrate in the diet results in utilization of fat for energy. In severe deficiency, incomplete oxidation of fats causes ketone bodies to accumulate in the blood.

Excess Carbohydrates

1. Excessive consumption of refined sugars could be one of the causes of den-tal caries or tooth decay.
2. Excessive sugar depresses the appetite, provides hollow calories, and could result in malnutrition.
3. High intake of sugar and refined carbohydrates increase the blood triglyceride levels leading to heart diseases.

4. When excessive carbohydrates are consumed they are converted into fat and deposited in the adipose tissue, which could lead to obesity, i.e., body weight of 20% or more than desirable weight.
5. Excessive fibre could irritate the intestinal lining causing cramps or bloating due to gas formation.
6. Excessive fibre interferes with the absorption and availability of mineral elements such as iron and calcium.

FAT

Fats consist of a wide group of compounds that are generally soluble in organic solvents and generally insoluble in water. Chemically, fats are triglycerides, triesters of glycerol and any of several fatty acids. Fats may be either solid or liquid at room temperature, depending on their structure and composition. Although the words "oils", "fats", and "lipids" are all used to refer to fats, "oils" is usually used to refer to fats that are liquids at normal room temperature, while "fats" is usually used to refer to fats that are solids at normal room temperature. "Lipids" is used to refer to both liquid and solid fats, along with other related substances, usually in a medical or biochemical context. The word "oil" is also used for any substance that does not mix with water and has a greasy feel, such as petroleum (or crude oil), heating oil, and essential oils, regardless of its chemical structure.

Fats form a category of lipid, distinguished from other lipids by their chemical structure and physical properties. This category of molecules is important for many forms of life, serving both structural and metabolic functions. They are an important part of the diet of most heterotrophs (including humans). Fats or lipids are broken down in the body by enzymes called lipases produced in the pancreas.

Examples of edible animal fats are lard, fish oil, butter/ghee and whale blubber. They are obtained from fats in the milk and meat, as well as from under the skin, of an animal. Examples of edible plant fats include peanut, soya bean, sunflower, sesame, coconut and olive oils, and cocoa butter. Vegetable shortening, used mainly for baking, and margarine, used in baking and as a spread, can be derived from the above oils by hydrogenation.

These examples of fats can be categorized into saturated fats and unsaturated fats. Unsaturated fats can be further divided into cis fats, which are the most common in nature, and trans fats, which are rare in nature but present in partially-hydrogenated vegetable oils.

Chemical Structure

There are many different kinds of fats, but each is a variation on the same chemical structure. All fats are derivatives of fatty acids and glycerol. The molecules are called triglycerides, which are triesters of glycerol (an ester being the molecule formed from the reaction of the carboxylic acid and an organic alcohol). As a simple visual illustration, if the kinks and angles of these chains were straightened out, the molecule would have the shape of a capital letter E. The fatty acids would each be a horizontal line; the glycerol "backbone" would be the vertical line that joins the horizontal lines. Fats therefore have "ester" bonds.

The properties of any specific fat molecule depend on the particular fatty acids that constitute it. Different fatty acids are composed of different numbers of carbon and hydrogen atoms. The carbon atoms, each bonded to two neighboring carbon atoms, form a zigzagging chain; the more carbon atoms there are in any fatty acid, the longer its chain will be. Fatty acids with long chains are more susceptible to intermolecular forces of attraction (in this case, van der Waals forces), raising its melting point. Long chains also yield more energy per molecule when metabolized.

Saturated and Unsaturated Fats

A fat's constituent fatty acids may also differ in the C/H ratio. When all three fatty acids have the formula $C_nH_{(2n+1)}CO_2H$, the resulting fat is called "saturated". Values of n usually range from 13 to 17. Each carbon atom in the chain is saturated with hydrogen, meaning they are bonded to as many hydrogens as possible. In unsaturated fats are derived from fatty acids with the formula $C_nH_{(2n-1)}CO_2H$. These fatty acids contain double bonds within carbon chain. This results in an "unsaturated" fatty acid. More specifically, it would be a monounsaturated fatty acid. A polyunsaturated fatty acid would be a fatty acid with more than one double bond; they have the formula, $CnH_{(2n-3)}CO_2H$. and $C_nH_{(2n-5)}CO_2H$. Unsaturated fats can be converted to saturated ones by the process of hydrogenation. This technology underpinned the development of margerine. Saturated and unsaturated fats differ in their energy content and melting point. Since an unsaturated fat contains fewer carbon-hydrogen bonds than a saturated fat with the same number of carbon atoms, unsaturated fats will yield slightly less energy during metabolism than saturated fats with the same number of carbon atoms. Saturated fats can stack themselves in a closely packed arrangement, so they can freeze easily and are typically solid at room temperature. For example, animal fats tallow and lard are high in saturated fatty acid content and are solids. Olive and linseed oils on the other hand are highly unsaturated and are oily.

Trans Fats

There are two ways the double bond may be arranged: the isomer with both parts of the chain on the same side of the double bond (the *cis*-isomer), or the isomer with the parts of the chain on opposite sides of the double bond (the *trans*-isomer). Most *trans*-isomer fats (commonly called trans fats) are commercially produced. Trans fatty acids are rare in nature. The *cis*-isomer introduces a kink into the molecule that prevents the fats from stacking efficiently as in the case of fats with saturated chains. This decreases intermolecular forces between the fat molecules, making it more difficult for unsaturated cis-fats to freeze; they are typically liquid at room temperature. Trans fats may still stack like saturated fats, and are not as susceptible to metabolization as other fats. Trans fats may significantly increase the risk of coronary heart disease.

Importance for Living Organisms

Vitamins A, D, E, and K are fat-soluble, meaning they can only be digested, absorbed, and transported in conjunction with fats. Fats are also sources of essential fatty acids, an important dietary requirement. Fats play a vital role in maintaining healthy skin and hair, insulating body organs against shock, maintaining body temperature, and promoting healthy cell function.

Fats also serve as energy stores for the body, containing about 37.8 kilojoules (9 calories) per gram of fat. They are broken down in the body to release glycerol and free fatty acids. The glycerol can be converted to glucose by the liver and thus used as a source of energy.

Fat also serves as a useful buffer towards a host of diseases. When a particular substance, whether chemical or biotic—reaches unsafe levels in the bloodstream, the body can effectively dilute—or at least maintain equilibrium of—the offending substances by storing it in new fat tissue. This helps to protect vital organs, until such time as the offending substances can be metabolized and/or removed from the body by such means as excretion, urination, accidental or intentional bloodletting, sebum excretion, and hair growth.

While it is nearly impossible to remove fat completely from the diet, it would also be unhealthy to do so. Some fatty acids are essential nutrients, meaning that they can't be produced in the body from other compounds and need to be consumed in small amounts. All other fats required by the body are non-essential and can be produced in the body from other compounds.

Adipose Tissue

In animals, adipose, or fatty tissue is the body's means of storing metabolic energy over extended periods of time. Depending on current physiological conditions, adipocytes store fat derived from the diet and liver metabolism or degrade stored fat to supply fatty acids and glycerol to the circulation.

These metabolic activities are regulated by several hormones (i.e., insulin, glucagon and epinephrine). The location of the tissue determines its metabolic profile: "visceral fat" is located within the abdominal wall (i.e., beneath the wall of abdominal muscle) whereas "subcutaneous fat" is located beneath the skin (and includes fat that is located in the abdominal area beneath the skin but *above* the abdominal muscle wall). Visceral fat was recently discovered to be a significant producer of signaling chemicals (i.e., hormones), among which are several which are involved in inflammatory tissue responses. One of these is resistin which has been linked to obesity, insulin resistance, and Type 2 diabetes. This latter result is currently controversial, and there have been reputable studies supporting all sides on the issue.

PROTEINS

Protein is the basic material of every living cell and is the most important of all known substances in the organic kingdom. It is the only nutrient that can make new cells and rebuild tissues. Therefore, an adequate amount of protein in the diet is essential for normal growth and development and for the maintenance of health.

Definition

Proteins are large, complex, organic compounds made up of carbon, hydrogen, oxygen, and nitrogen. The presence of nitrogen distinguishes proteins from carbohydrates and fats. Apart from nitrogen, elements such as sulphur, phosphorus, copper, and iron are also found in some proteins. The basic units from which proteins are built are the amino acids. Each amino acid contains a carboxyl group (COOH) or acid group and an amino group (NHS or basic group. Proteins consist of chains of amino acids that are linked to each other by a pep. tide linkage (-CO-NH-). Twenty-two different amino acids are widely distributed in nature.

However, the proteins obtained from plants and animals are quite different both in amounts present and in quality. The proteins that make up the skin, bones, muscles, hair, and nails in the body are obviously very different from each other. Egg, milk, meat, and pulse proteins also differ both in quality and in quantity. This is because the amino acids present in each of these proteins are in different permutations and combinations. Thousands of different proteins exist in nature and vary widely from one another in quality. No two proteins will have an identical amino acid content. The protein content of any food can be estimated by measuring the nitrogen content of the food. Since proteins contain 16% nitrogen, each gram of nitrogen measured is equal to 6.25 g of protein.

Essential Amino Acids

Those amino acids which cannot be synthesized in sufficient amounts by the body and must be provided by the diet are called essential amino acids. The human adult requires eight essential amino acids, while growing children require ten essential amino acids. Essential amino acids are indispensable to life.

Non-essential Amino Acids

All amino acids are required by the body for tissue synthesis and repair. Non-essential amino acids does not mean that these amino acids are not required by the body. They are termed non-essential because they are not dietary essentials. If they are lacking in the diet, they can be synthesized by the body from other amino acids. The twenty-two amino acids present in proteins could be compared with the letters of the Roman alphabet and the innumerable words present in our dictionary. Similarly, innumerable proteins can be formed by using the twenty-two amino acids in varying sequences and quantities.

PROTEIN QUALITY

No two food proteins are identical in their quality, i.e., with the efficiency with which they can be used in the body. The quality of protein depends upon the kind and amount of amino acids present in them in relation to the body needs. Protein quality is an important criterion for tissue synthesis. Each body protein performs a specific function and cannot be replaced by another protein. When a new protein has to be synthesized, all the amino acids, which make up the protein, must be availableat the same time and in sufficient quantity. Even if one amino acid is missing or deficient, protein cannot be synthesized just like a word cannot be made if even one of the alphabets of which it is made up of is missing.

Biological Value

It is an index of protein quality. It is defined as the amount of absorbed nitrogen retained in the body. The digestibility factor is not taken into account. Biological value (BV) is a quantitative measure of the nutritive value of a protein food. A pro-tein of high BV will retain more nitrogen than a protein of low BV. Cereals and pulses consumed together will have a higher BV than the average value of the individual cereal or pulse. This is because of the complementary nature of proteins. Amino acids deficient in cereals will be compensated by the amino acids present in pulses.

CLASSIFICATION OF PROTEINS

Proteins may be classified on (1) the basis of their structure or on (2) the basis of their quality, i.e., the amino acids present in them.

Classification by Structure

Simple proteins These proteins are made up of amino acids only, e.g., zein in corn, albumin in egg white, and gliadin in wheat consist of amino acids only.

Conjugated proteins These proteins have a nonprotein molecule attached to the protein, e.g., blood protein haemoglobin, which contains a haeme (iron) group attached to protein and milk protein casein, which has a phosphate group attached. Derived proteins These result from a partial breakdown of a native protein.

Proteoses, peptones, and polypeptides are formed when digestive enzymes begin their action on proteins. From the nutritional point of view, classification of proteins on the basis of their quality is more relevant than classification by structure.

Classification by Quality

Proteins are classified into three groups on the basis of their quality. Complete proteins These proteins contain all essential amino acids in sufficient proportions and amounts to meet the body's need for growth and repair of tissue cells.

A complete protein food has a high BV Eggs, milk, meat, fish, and poultry are complete protein foods. They are found in animal foods. Partially complete proteins These are proteins in which one or more essential amino acids are present in inadequate amounts.

They cannot synthesize tissues without the help of other proteins. The value of each is increased when it is consumed in combination with another incomplete protein at the same meal.

They can maintain life. They are found in plant foods. Cereals, pulses, nuts, and oilseeds are partially complete protein foods. Cereals contain inadequate amounts of essential amino acid lysine, and pulses are deficient in essential amino acid methionine.

Incomplete proteins: These proteins are incapable of growth and repair of body cells. They cannot maintain life. One or more essential amino acids may be completely lacking in these proteins, e.g., gelatin and zein in corn. Gelatin lacks three essential amino acids and is the only animal protein which is incomplete. Proteins perform three main functions - structural function, regulatory function, and energy.

Structural Function

Growth The primary function of food protein is the synthesis of body cells. All body tissues and fluids except urine and bile are made up of protein. Proteins are the major constituent of muscles, organs, endocrine glands, and collagen. Collagen is the main structural protein of bones, tendons, ligaments, skin, blood vessels, and connective tissue. All enzymes and some hormones, e.g., insulin are made up of proteins. Proteins are required for the formation and growth of all these substances. During periods of rapid growth, additional proteins are needed for synthesis of body components.

Energy

Maintenance or wear and tear Protein is required by all age groups for continuous maintenance of all the cells in the body. Cells have a varying lifespan and proteins are needed to replace the old or worn out cells.

Regulatory Functions

All amino acids from food protein are used for growth and maintenance. Certain amino acids and proteins have highly specialized functions in the regulation of body processes and protection against disease. Some of the regulatory functions are as follows.

1. Haemoglobin, an iron containing protein in the red blood cells, performs an important role by transporting oxygen to the tissue cells.
2. Plasma proteins maintain water balance and regulate the osmotic pressure in the body.
3. Antibodies that are protein in nature perform a protective function by increasing the body's resistance to disease.
4. All enzymes and some hormones, e.g., insulin are made up of protein. The hormone insulin regulates blood sugar levels. Enzymes act as specific catalysts to metabolic processes in the body.
5. Some amino acids have specific functions, e.g., tryptophan serves as a precursor for niacin, a B-complex vitamin. The amino acid tyrosine in combination with iodine forms the hormone thyroxine.

Like carbohydrates, proteins too provide 4 kcal/g when broken down in the body. The basic need of the body is energy and this takes priority over protein synthesis. If the diet does not supply adequate calories from carbohydrates and fats, the proteins from the diet will be oxidized to meet the energy needs of the body.

If the diet is deficient in calories, the body uses up its protein and fat stores. Using protein as a source of energy is not advisable as it puts an extra burden on the body and the pocket. Protein is used by the body as a source of energy only when no other source of energy is available.

SPECIFIC ELEMENTS OF SOME HIGH-PROTEIN DIETS

The most popular of all high-protein diets is the Atkins Diet introduced by Robert C. Atkins, a New York City cardiologist, in his 1970s best-seller Dr. Atkins' Diet Revolution. Since that first book on his diet, Atkins has written several more. His works have sold millions of copies.

Atkins contends that the majority of people follow the wrong route to weight loss. They eliminate red meat from their diets and prepare their scrambled egg whites with little or no oil; they remove the skin from their chicken and use no butter on their potatoes and pasta. Oatmeal and skim milk become preferred breakfast foods, and desserts may consist of frozen yogurt or fruit sorbet. But the weight does not drop. Atkins says that most doctors believe that the only way to lose weight is to reduce calories. To Atkins, that is incorrect. "Different kinds of diets can have different effects on the amount of calories a person's body consumes daily, and, by taking different metabolic pathways, can cause the body to require different amounts of energy to do its work. On a low-carbohydrate diet, there are metabolic advantages that will allow you to eat as many or more calories as you were eating before starting the diet and still begin losing pounds and inches. And if you eat fewer calories-most people do on this diet-you'll lose weight very fast" (Atkins: 10).

There are negative aspects that Atkins does not address. High amounts of dietary cholesterol-especially over an extended period of time-are dangerous. Foods high in cholesterol contribute to an elevated cholesterol level and other cardiovascular problems. While rapid weight loss may be psychologically desirable, it is not necessarily good for the body. It is far healthier to lose weight continuously over a longer period of time. Atkins says that Americans historically ate very little sugar. In 1828, the average American consumed only 12 pounds a year. In 1990, it was found that the typical American annually ate 137.5 pounds of sugar (Atkins: 44). "Since the average person eats 425 grams of carbohydrates a day, 105 grams of protein, and 168 grams of fat, sugar constitutes 40% of the carbohydrate total and nearly one fourth of our daily food by weight. Sugar has no nutritional value and is directly harmful to your health" (Atkins: 45).

Atkins says that earlier Americans were carnivores whose diets included beef, fish, fowl, pork, and eggs. But, he maintains, hardly anyone died from heart attacks. As a result, Atkins claims that the human body was designed to eat those types of foods. He says that the host of medical problems facing Americans today is largely caused by the contemporary diet. Still, the lifestyle of the average earlier American required a good deal of hard labor, and they probably burned off much of the fat that they ate. The Atkins diet does not focus on exercise. Yet, Atkins does advise it and calls it an "essential part of the program" (Atkins: 202). Atkins says that the amount of insulin in the blood is a function of the level of carbohydrates. Diets that are high in carbohydrates encourage the production of insulin. When there is too much sugar in the blood for the available insulin to handle, the excess is converted to fat. When fewer carbohydrates are consumed, the body will burn its own fat for energy. During the first 14 days of the Atkins diet, known as the "Induction Diet," dieters are able to eat all the high-protein foods they wish. Carbohydrates are limited to 20 grams per day-"about one hot-dog bun's worth." The goal is to "trigger ketosis," the burning of stored fat for energy. Then, until the ideal weight is achieved, between 25 and 75 grams of carbohydrates may be consumed each day. When maintenance is reached, there may be a little more flexibility (Jibrin: 78).

In Dr. Atkins' New Diet Revolution, Atkins describes people who have witnessed dramatic improvements on his diet. One example is an overweight man in his 60s who survived three heart attacks. The man also had arthritis and high cholesterol. Atkins eliminated ice cream and French fries from his diet and "urged him to eat a lot of meat, fish, fowl, and eggs when he wanted them, nuts, salads, vegetables, and a little cheese." The man's cholesterol went down from 228 to 157, and his weight dropped from 228 to 190, "a pretty good weight for a powerfully built six-footer." The man also experienced an improvement in his arthritis symptoms (Atkins: 10-11).

Another hugely popular high-protein diet, the Zone, was introduced by Barry Sears, a biochemist, in his mid-1990s best-seller The Zone. In the preface of this book, Sears explains his motivation to find a healthier lifestyle: he was a "walking genetic time bomb" (Sears, 1995: ix). Sears says that he was genetically programmed to die young of heart disease. Before they reached the age of 54, his grandfather, father, and three uncles died from heart attacks. Searching for a way to avoid this fate, Sears eventually came upon eicosanoids-hormones that "act as 'master switches' that control virtually all human body functions-including the cardiovascular system, the immune system, and the systems that govern how much fat we store (and therefore how much we weigh)." Sears believes that many diseases such as heart disease, diabetes, arthritis, and cancer are caused by imbalances among the eicosanoid hormones. He claims that maintaining and restoring the balance "might help prevent or even become the primary treatment for these diseases" (Sears, 1995: xiv).

Eicosanoids have the potential to do even more. They "might help maintain a state of nearly perpetual good health: a molecular definition of 'wellness' that would lead to a better quality of life. As an ultimate payoff, keeping eicosanoids in balance might help all of us reach that near-euphoric state of maximum physical, mental, and psychological performance that athletes call 'the Zone' " (Sears, 1995: xiv). Sears says that his diet uses food "to manipulate

[the] eicosanoid balance." This manipulation is a "passport to the Zone" (Sears, 1995: xv). In Sears's system, all calories are not created equal. "The hormonal effect of a calorie of carbohydrate is different than the hormonal effect of a calorie of protein, and is still different from the hormonal effect of a calorie of fat. Each of these three nutrients has its own unique effects on your body's hormones. In the proper balance, these three nutrients are exactly what your body needs to remain healthy by keeping insulin within the Zone." When the nutrients are imbalanced and insulin levels rise, the body's hormonal stability is altered, "resulting in weight gain, an increased likelihood of chronic disease, and acceleration of the aging process." However, "if insulin levels are too low, your cells begin to starve because a certain amount of insulin is required to drive life-sustaining nutrients into your cells" (Sears, 2000:3-4). Sears recommends a diet that is 40% carbohydrate, 30% protein and 30% fat; it has a daily intake of between 1,100 and 1,600 calories. Each meal should include a low-fat protein. In the Zone diet, protein is obtained from leaner sources such as turkey, fish, skinless chicken, egg whites, soy meat substitutes, tofu, and 1% cottage cheese. A tiny amount of good fat, such as that in nuts or olive oil, is advised. At least for the first two weeks, rather than eating pasta, bread, and potatoes, carbohydrates should be obtained from fruits and vegetables. According to Sears, people who follow his diet should lose between 1 and 1 1/2 pounds per week without feeling hungry (Stedman: 73-74). After The Zone, Sears wrote several more books. In A Week in the Zone he summarized the basic Zone rules:

1. Always eat a Zone meal within one hour after waking.
2. Try to eat five times per day: three Zone meals and two Zone snacks.
3. Never let more than five hours go by without eating a Zone meal or snack.
4. Eat more fruits and vegetables and less bread, grains, and other starches.
5. Drink at least eight eight-ounce glasses of water every day.
6. Don't worry about an occasional mistake at a meal; just be sure to make the next meal a Zone meal. (Sears, 2000:5-6)

In the same book, Sears listed some of the long-term benefits of the Zone diet: permanent fat loss, reduced risk of heart disease, breast cancer, and type 2 diabetes, protection from arthritis and osteoporosis, and fewer infections (Sears, 2000:19-20). He also disputed the generally accepted notion that his diet is high in protein, noting that it consists of more carbohydrates than protein (Sears, 2000:16-17). In The Carbohydrate Addict's Lifespan Program, Richard F. Heller and Rachael F. Heller maintain that some people are "addicted" to carbohydrates. They define this addiction as "a compelling, recurring or, at times, escalating craving for starches such as bread, pasta, rice, or potatoes, or for snack foods such as chips, popcorn, or pretzels, or for sweets such as cookies, cakes, pies, donuts, muffins, or chocolate." These people have "a tendency to gain weight easily or, over time, to regain weight that has been lost by dieting." The Hellers add that carbohydrate addicts do not necessarily eat more food than nonaddicts. Sometimes they eat less, but, "The carbohydrate addict's body...may be much more efficient at turning food energy into fat".

According to the Hellers, carbohydrate addiction is caused by an imbalance of insulin in the body. Some carbohydrate addicts produce too much insulin. Others have cells that resist insulin, and the excess insulin remains in the bloodstream. Some carbohydrate addicts have both problems (Heller and Heller: 24). Whatever the cause, there is too much insulin in the blood.

In essence, the Hellers advise an eating plan that has two high-protein/ high-fibre vegetable meals per day and one meal that includes carbohydrates. The meal with carbohydrates should begin with two cups of fresh salad. The remaining foods should look like a plate divided into thirds. "Place the vegetables, the proteins, and the carbohydrates (including dessert) in each of the thirds" (Heller and Heller: 112). Second helpings are permitted, but they must be eaten in the same proportions-no seconds of carbohydrates without the vegetables and proteins. All the food should be consumed within an hour. A daily consumption of six to eight glasses of water is required.

In the hugely popular Sugar Busters! H. Leighton Steward, a writer and former business executive, and doctors Morrison C. Bethea, Samuel S. Andrews, and Luis A. Balart contend that the primary culprit for ill health and excess weight is sugar. In their diet, people need to avoid "foods or combinations of foods that require the secretion of large amounts of insulin." To these authors, "refined sugar in any significant quantity is toxic to many human bodies, and it certainly helps make many bodies fat" (Steward et al.: 2-3).

They advise a diet of "natural unrefined sugars, whole unprocessed grains, vegetables, fruits, lean meats, fibre, and alcohol (in moderation)" and provide lists of acceptable foods (Steward et al.: 118, 123-126).

Foods such as potatoes, beets, and carrots, which are "simply starch, a storage form of glucose," are eliminated because they are converted to pure sugar during digestion (Steward et al.: 120-121). Sugar Busters! advises drinking only a small amount of fluids with meals. It contends that excess fluids impair digestion. On the other hand, additional fluids may be consumed between meals.

LIMITATIONS OF HIGH-PROTEIN DIETS

A 1997 article in the Harvard Health Letter raised a number of concerns about high-protein diets. While it is true that about 25% of Americans are insulin resistant, it is generally believed that this is not caused by eating too many carbohydrates. On the contrary, it is thought to be a direct result of excess weight and a sedentary lifestyle. Some people lose a considerable amount of weight on high-protein diets. When they do, this is assumed to be caused by eating fewer calories. The American Dietetic Association suggests that a 154-pound man eats about 2,600 calories per day. A man of that size on the Zone diet would eat about 900 calories a day. That is quite restrictive (Norris: 1-3).

The same article notes that Americans do not consume inadequate amounts of protein. On the contrary, Americans eat about 80 to 90 grams each day, almost twice what they need. In sharp contrast to the proponents of high-protein diets, this article emphasizes that these diets foster cardiovascular disease and high cholesterol levels and increase the risk of cancer. The article also criticizes Atkins's view of ketosis, or the burning of fat for energy.

Because of the loss of water, sodium, and potassium, ketosis may trigger an uneasy feeling and lightheadedness. The lack of fibre may result in constipation. There may be a diminished appetite and nausea. If the uric acid level rises too high, it could set off an episode of gout (Norris: 1-3).

While acknowledging that people are losing weight on high-protein diets, a 1999 story in Men's Health mentioned some of the limitations. Like the previous article, it expressed concern about ketosis, which may also produce bad breath and heart irregularities. It further noted that there might be a reduced ability to exercise. The diet could also be dangerous for people with impaired kidney function (Jibrin: 78).

A 2000 article in the Journal of the American Dietetic Association said that books on high-protein diets are filled with anecdotal evidence, but there are few if any hard data to prove the theses. "Often the information communicated in the diet books is exaggerated or interpreted in such a way that cannot be proven" (Stein: 760).

Additionally, the International Food Information Council says, "There is a dire lack of scientific research to corroborate the theories expounded in the majority of diet books currently on the market.... Authors may simplify or expand upon biochemistry and physiology in an effort to help support their theories and provide a plethora of scientific jargon that people do not understand but that seems to make sense. And few, if any, offer solid scientific support for their claims in the form of published research studies. Instead, most evidence is based on anecdotal findings, theories, and testimonials of short-term results." This Council also lists concerns specific to each of several high-protein diets, such as those of Atkins and Sears.

Writing in 1999 in Prevention, Elizabeth Somer, a registered dietician, says that high-protein diets may prompt depression. "Ironically, all that protein has the effect of lowering the level of the amino acid tryptophan in your brain, which is the building block for the mood-lifting brain chemical serotonin. High levels of serotonin boost your mood, but low levels can result in depression. No other nerve chemical is as strongly linked to your diet as is serotonin." In sharp contrast to the high-protein advocates, Somer advises people to "make sure that every meal contains some carbohydrate-rich foods, especially whole grain foods such as whole wheat bread, oatmeal, or brown rice. Carbohydrates trigger the release of insulin, which allows tryptophan to freely enter your brain, causing serotonin levels to rise" (Somer, December 1999:124-129).

Two separate studies published in 1998 in the American Journal of Clinical Nutrition found that high-protein diets may predispose people to some nutritional deficiencies. One study found that a high-animal-protein diet was correlated with the loss of calcium in the urine (Itoh, Nishiyama, and Suyama: 438-444). The other study determined that young women on high-protein diets appear to have greater requirements for vitamin B6 (Huang et al.: 208-220).

A 2000 article in Runner's World advises runners not to follow a high-protein diet for a number of reasons. The diet does not contain a sufficient amount of carbohydrates to "fuel" running. Runners require a minimum of 400 grams per day of carbohydrates. "But possibly the worst part of the high-protein diets-for runners-is how tired you feel.... After a week or so on the diet, you must either give up running or agree that this low-carbohydrate thing is for the birds (or more accurately, cats, who survive best on a high-protein diet)" (Applegate: 30). A 1997 article in Prevention magazine included input from Jay Kenny, a nutritionist with the Pritikin Longevity Center. According to the author, Colleen Pierre, Kenny wondered why the Japanese, who eat a high-carbohydrate diet, are "enviably slim." Insulin, he added, does not make people overweight. Instead, extra weight raises insulin levels. Losing weight generally reduces the amount of insulin in the blood (Pierre: 85-91). A study published in 1998 in Pediatrics described the effects of a high-protein, very low-calorie diet on six morbidly obese adolescents. The adolescents, between the ages of 12 and 15, all weighed more than 200% of the ideal body weight. For the first eight weeks, "daily intake consisted of 650 to 725 calories, which was substantially in the form of protein (80-100g).

The diet was very low in carbohydrates (25g)." During this phase, the patients lost an average of 15.4 kg, an amount that was statistically significant. Then, for an additional 12 weeks, two carbohydrates (30g) were added at each meal. While on this phase of the diet, one subject regained the weight. The remaining five lost an average of 4.9kg, a figure that was not statistically significant. As a result, the researchers concluded that the very low-calorie, high-protein diet "can be used effectively for rapid weight loss in adolescents with morbid obesity. Loss in lean body mass is blunted, blood chemistries remain normal and sleep abnormalities significantly decrease with weight loss." Nevertheless, there were a few words of caution. Although the subjects consumed large amounts of supplemental calcium, during the high-protein phase, "bone mineral density decreased significantly" (Willi et al.: 61-67).

Though it does not advocate high-protein diets, a 2000 article in Prevention lists some ways to make them healthier:

- Have your doctor test your kidney function before you start a high-protein diet to make sure that your kidneys are healthy.
- Choose a diet that maximizes intake of high-fibre, brightly colored vegetables and be sure to eat all the vegetables that are allowed.
- Select a diet that emphasizes low-saturated-fat meat and poultry. Trim fat. Don't char.
- Pick a diet that includes at least 50 grams of carbohydrates daily to prevent ketosis.
- Drink 8 cups of fluid daily to help the kidneys.
- Select whole-grain breads and cereals when possible.
- Choose soy-based foods, legumes, and fish to provide protein.
- Choose low-fat dairy foods and calcium supplements to reach 1,500 to 2,000 milligrams of calcium daily.
- Take a standard (not high-potency) multivitamin/mineral supplement daily. (Pierre and Robertson: 134-141)

4

The Vitamins

The term vitamin is derived from the Latin words 'vital' and 'amine', because vitamins are required for life and were originally thought to be amines. An organic compound is considered a vitamin if a lack of that compound in the diet results in overt symptoms of deficiency.

Vitamins are classified as either water-soluble (all B vitamins and C) or fat-soluble (A, D, E, K). While fat-soluble vitamins are stored in the body's fatty tissue, water-soluble vitamins must be used by the body right away. Any left over water-soluble vitamins leave the body through the urine. Vitamin B12 is the only water-soluble vitamin that can be stored in the liver for many years.

Vitamins are essential to life and healthy living. Failing to get the necessary amounts of specific vitamins can cause deficiency states that are unhealthy and even dangerous. Thus, a sufficient intake of vitamins is crucial to prevent the development of deficiency-related diseases. In addition, some vitamins have a considerable potential in health promotion and disease treatment.

Animal diets are produced from raw materials (grains, soybeans, etc). These raw materials most of the time do not contain sufficient vitamins to cover the needs of pigs, poultry, ruminants or pets. Extra vitamins are added to the diet to cover these needs. Vitamins can be added as single ingredient or as a premix (mixture of several vitamins). Orffa can offer a broad and complete range of vitamins to the compound feed and premix industry.

Vitamins

Vitamins are closely linked to mineral metabolism and absorption. Vitamin A helps skin and mucous membranes stay healthy. Vitamin A requirements usually are met by grazing fresh, green, growing grass. Oxidation deteriorates vitamin A during storage, so diets based on stored feeds should be supplemented with vitamin A. Supplement diets with vitamin A any time the major portion is stored feeds.

Vitamin A can be added to a mineral mix in a stabilized form to prevent oxidation. The minimum amount should be approximately 120,000 International Units (IU) of vitamin A per pound of mineral.

Vitamin A can also be added to the grain mixture to provide 15,000 to 30,000 IU per head per day, depending on individual requirements. An alternative method is to inject 1.5 million IU subcutaneously if a source of dietary citamin A is not available for 60 to 90 days, although unnecessary injections are discouraged in consideration of National Beef Quality Assurance guidelines.

Vitamin D aids the absorption of calcium and phosphorus from the intestine and their deposition in the bone matrix. Signs of vitamin D deficiency are similar to a calcium or phosphorus deficiency. Most cattle exposed to direct sunlight synthesize enough vitamin D, but cattle in a covered confinement feedlot may need supplemental vitamin D.

Vitamin E is usually present in the diet in sufficient quantities for all classes of cattle; however, a selenium deficiency could lead to an apparent deficiency of vitamin E. Vitamin E can be helpful for short-term periods of stress that may occur when calves are co-mingled and transported at weaning.

Other essential vitamins are usually present in adequate quantities in the diet or are synthesized by bacteria in the rumen.

FAT-SOLUBLE VITAMINS

Fat-soluble vitamins A, D, E, and K differ from the water-soluble vitamins in their accumulative action.

Little evidence has been recorded for hypervitaminosis with the water-soluble vitamins since these compounds are rapidly metabolized and excreted when intake exceeds liver or tissue storage capacity, but hypervitaminosis is a common occurrence in fish and other animals when large quantities of any one of the fat-soluble vitamins are ingested.

Toxicity symptoms involving vitamins A and D are indistingui-shable from deficiency symptoms for the same vitamins.

On the other hand, symptoms of excess vitamins E and K intake are more discrete. Fish rations may often be enriched with fish oils to increase caloric density of the ration resulting in excessive intake of the fat-soluble vitamins.

Vitamin A

A fat-soluble factor that promoted growth in rats was described by Hopkins and by Osborne and Mendel at the beginning of 20th century.

McCollum and Simmonds cured xerophthalmia, a disease of the eye, with this material.

The chemical structure of vitamin A and its relationship to -carotene was shown by Von Euler in 1928. Active vitamin A was synthesized in the mid-thirties.

Positive Functions

Vitamin A is essential in maintaining epithelial cells, preventing atrophy and keratinization of epithelial tissues. It is combined with a protein in visual purple and is important in night vision. Vitamin A also prevents xerophthalmia in rats and young children. Vitamin A promotes growth of new cells and aids in maintaining resistance to infection.

The Syndrome of Deficiency or Excess

Hypovitaminosis A is characterized by poor growth, poor vision, keratinization of epithelial tissue, xerophthalmia, night blindness, haemorrhage in the anterior chamber of the eye, haemorrhage of the base of the fins, and abnormal bone formation. Nerve degeneration has been reported in pigs, chickens, rats, rabbits, and ducks, but only occasionally observed in fish after long periods of deficiency. Hypervitaminosis A has been described in fish and in other animals and involved enlargement of liver and spleen, abnormal growth, skin lesions, epithelial keratinization, hyperplasia.

Of head cartilage, and abnormal bone formation resulting in ankylosis and fusion of vertebrae. Hypervitaminosis A is reflected in very high liver oil vitamin A content and elevated serum alkaline phosphatase. Removal of excess vitamin A from the diet promotes rapid recovery.

Requirements

Dietary requirements for vitamin A alcohol are listed in Table Fish apparently require vitamin A for growth in light, but not in darkness. Therefore, the requirement for maximum growth and reproduction is related to exposure to light and reflects observations in other animals that near normal growth will occur at very low vitamin A intake in protected environments where fish are not exposed to stress, infection, and ultraviolet radiation.

Sources and Protection

Cod liver oil is the best known source of vitamin A although black sea bass, swordfish, or ling cod oils contain much more of the vitamin. Whale liver oil contains kitol which has little or no biological activity until heated above 200° C when one molecule of biologically active vitamin A is generated per molecule of whale kitol.

This biologically inactive kitol may be deposited in the whale to avoid hypervitaminosis A during excessive vitamin A intake. The possibility of hypervitaminosis A occurs when tuna, shark, or ling cod viscera are used in preparation of moist fish diets. Synthetic vitamin A preparations, such as vitamin A palmitate, are available and are often used to supplement rations low in fish meal, fish viscera, or carotenes. Some fish species seem able to utilize b-carotene as a vitamin A source, whereas others are unable to split the g-carotene molecule and vitamin A must be added to the diet.

Clinical Assessment

Vitamin A status can best be assessed by the absence of deficiency signs and by assay of liver oil for vitamin A content. Assay for the vitamin in blood or plasma has not been found to be useful.

Vitamin D

Rickets was induced with test diets by Hopkins, and Mellanby cured the disease in dogs by adding cod liver oil to the ration. Ultraviolet light exposure was shown to have an anti-rachitic effect, and provitamin D activity had been ascribed to ergosterol. Crystalline vitamin D was isolated by Angus and activated 7-dehydrocholesterol was isolated by Windaus.

Positive Functions

Vitamin D is essential for maintaining calcium and inorganic phosphate homeostasis. It is involved in alkaline phosphatase activity, promotes intestinal absorption of calcium, and influences the action of parathyroid hormone on bone. Fish may sequester calcium from water through the gill membrane; thus the major function of vitamin D for other animals may not be necessary to satisfy calcium requirements for fish.

Syndrome of Deficiency or Excess

Hypovitaminosis D for fish has not been described. Rickets and abnormal bone formation described in detail for animals has not been observed in fish fed low vitamin D diets. However little work has been done under carefully controlled conditions with young growing fish and only alteration in alkaline phosphatase activity has been reported on different vitamin D diet intake. Hypervitaminosis D, however, has been reported. Brook trout fed large doses of vitamin D showed impaired growth, lethargy, and dark colouration. High intake of vitamin D mobilizes phosphorus and calcium from the bone and tissues and may result in fragile bones, poor growth, and poor appetite related to the nausea described in man afflicted with hyper vitaminosis D. Arterial and kidney lesions due to excessive intake of vitamin D reported for rats and dogs have not been described histologically for fish nor has hypercalcaemia been described. This area needs to be explored because of the potential for hypervitaminosis D in fish fed diets containing various fish viscera which might contain large amounts of the vitamin D. Tuna liver oil may contain, for example, 100 to 1 000 times as much active vitamin D as cod liver oil.

Requirements

No requirements of fish for vitamin D have been described to date, but little work has been done in this area using highly purified diets and controlled experiments with young growing fish. Likewise, no demonstration of synthesis of the vitamin in fish reared on diets devoid of vitamin D or vitamin D precursors has been reported. Therefore, actual requirement of vitamin D for maintenance of calcium-phosphorus homeostasis in the young growing animal has not been adequately investigated and the true requirements of vitamin D in young fish have not been defined.

Sources and Protection

Vitamin D requirements of many animals can be met by ultraviolet irradiation of the skin where cholesterol derivatives are converted to calciferol.

Since the vitamin is fat soluble and accumulates in lipid stores, fish liver oil is a rich source of the material.

Content varies tremendously in liver oil, however, with values of about 25 I.U./g present in shark liver oil and over 200 000 I.U./g in albacore tuna liver oil. Cod liver oil contains from 100-500 I.U./g and animal liver contains some vitamin D. One international unit (I.U.) is equal to 0.025 mg of crystalline vitamin D.

Clinical Assessment

Absorption maxima in the ultraviolet region can be used to detect provitamins D in the non-saponifiable fraction of oils. Concentrated preparations of vitamin D can be assayed by the Carr-Price antimony trichloride reaction when assay is necessary to determine biologically active materials in liver oil of fish on different treatments.

The chick assay may not apply to fish liver storage levels because the biologically active form for fish has not been determined. Therefore, clinical assessment for hypervitaminosis A must rely on crude methods for determination of vitamin D in liver oil samples by the Carr-Price reaction or by measuring absorption in the UV spectrum.

Vitamin E

The existence of an antisterility vitamin was postulated by Evans and Bishop. This factor was named 'vitamin E' by Sure. The active tocopherol was isolated, characterized, and synthesized by Karrer.

Chemistry

Vitamin E is composed of a class of compounds known as tocopherols. One of the most important tocopherols is α-tocopherol. Eight naturally occurring tocopherol derivatives have been isolated and all belong to the D series. Synthetic -tocopherol is a racemic DL α-tocoperhol mixture. The derivatives of tocol or of tocotrienol are named, alpha, beta, gamma, delta, epsilon, eta, $zeta_1$ and $zeta_2$-tocopherol.

Table. The Tocopherols

αTocopherol (alpha)	5,7,8-trimethyltocol
βTocopherol (beta)	5,8-dimethyltocol
λTocopherol (gamma)	7,8-dimethyltocol
ζ_2-Tocopherol (zeta2)	5,7-dimethyltocol
η-Tocopherol (eta)	7-methyltocol
δ-Tocopherol (delta)	8-methyltocol
ε-Tocopherol (epsilon)	5,8-dimethyltocotrienol
ζ_1-Tocopherol (zeta1)	5,7,8-trimethyltocotrienol

The pure tocopherols are fat-soluble oils which are capable of esterification to form crystalline compounds. The tocopherols are stable to heat and acids in the absence of oxygen, but are rapidly oxidized in the presence of nascent oxygen, peroxides, or other oxidizing agents. The tocopherols are sensitive to ultraviolet light and are excellent antioxidants in the free form.

The esters are more stable and are commonly used as dietary supplements anticipating hydrolysis in the gut and absorption of the free alcohol to act as an active intra- and intercellular antioxidant. Ethyl derivatives on the aromatic ring are also active. Oxidation products of a-tocopherol can be reduced with hydrosulphite to -tocopheryl-hydroquinone or, in the presence of ascorbic acid, to α-tocopherol.

Positive Functions

The tocopherols act as extra and intracellular antioxidants to maintain homeostasis of labile metabolites in the cell and tissue plasma. As physiological antioxidants, these usually protect oxidizable vitamins and labile unsaturated fatty acids. Vitamin E prevents encephalo-malacia in chicks, erythrocyte haemolysis in several animals and steatitis in mink, pigs, and farm animals. The tocopherols also prevent exudative diathesis in chicks, white muscle disease in lambs and calves, and dietary liver necrosis in rats.

They act with selenium and with vitamin C for normal reproductive activity and are involved in prevention of nutritional muscular dystrophy in the chick, the yellowtail and carp. The tocopherols act as free radical traps to stop the chain reaction during peroxide formation and stabilize unsaturated carbon bonds of polyunsaturated fatty acids and other long-chain labile compounds. Vitamin E is involved in the maintenance of normal blood capillary permeability and the integrity of heart muscle. It was first shown to be involved in prevention of sterility and foetal resorption in rats. It may likewise affect embryo membrane permeability and hatch-ability of fish eggs.

Syndrome of Deficiency or Excess

One of the first signs is erythrocyte fragility closely followed by anaemia, ascites, xerophthalmia, poor growth, poor food conversion, epicarditis, and ceroid deposits in spleen and liver. Muscle dystrophy and xerophthalmia have been described in yellowtail and carp.

Non-specific forms of degenerative conditions have been described in several species of fish fed large quantities of polyunsaturated fatty acids with inadequate tocopherol in the ration. Hypervitaminosis E results in poor growth, toxic liver reaction, and death.

Requirements

The exact requirement of the fish for α-tocopherol may depend upon the amount and type of polyunsaturated fatty acids content of the ration. Polyunsaturated labile fish oils may invoke an increased requirement for intracellular antioxidants.

In addition, the amount of tocopherol needed as a supplement to the ration will depend upon the form of the vitamin used, the method of diet preparation, and the storage conditions before feeding.

Sources and Protection

Vegetable oils are rich sources of tocopherols. Synthetic -tocopherol in esterfied acetate or phosphate form is commonly used as a diet supplement. These esters are much more stable than the free form which is rapidly lost by air oxidation or in the presence of labile compounds like the polyunsaturated fish oils. Addition of antioxidants such as BHT (butylated hydroxyanisole) and BHT (butylated hydroxytoluene) protects fats and other labile compounds in the ration from oxidation, but these antioxidants have no vitamin E activity.

Clinical Assessment

The erythrocyte fragility test indicates the physiological state of fish. Absence of histologically detectable ceroid in liver and spleen from representative samples in the population is a good clue on the presence of adequate amounts of physiological antioxidants in the fish.

A barbituric acid test for oxidation of components has not been applied to fish tissues as a clinical tool for assessment of nutrition state except when liver oils become saturated with very labile polyunsaturated fatty acids such as when feeding squid or saury oils. These assays are good indicators of the state of oxidation of the finished ration, but the absence of deficiency signs and normal erythrocyte fragility are better clinical tests. Analysis for tocopherol is difficult and time-consuming and only applicable under critical research experiment situation.

Vitamin K

The name vitamin K (for 'koagulation') was proposed by Dam. Dam isolated the vitamin from alfalfa and from fish meal in 1939. It was synthesized later that year.

Positive Functions

Vitamin K is involved in the hepatic synthesis of blood clotting proteins - prothrombin and proconvertin. Substituted forms of vitamin K are strongly bacteriostatic and may serve as an alternate defence mechanism for bacterial infections. Vitamin K is structurally related to flavo compounds which function as electron carriers in the electron transport system. The primary role of vitamin K is to maintain a fast normal blood clotting rate which is so important to fish living in a water environment.

Syndrome of Deficiency or Excess

Prothrombin time in salmon fed diets devoid of vitamin K was increased three to five times and, during prolonged deficiency states, anaemia and haemorrhagic areas appeared in the gills, eyes, and vascular tissues. Increased blood clotting time has also been reported for other fish reared on diets with low vitamin K content. Interrelationships with other vitamins have not been documented in fish where the primary deficiency signs are slow blood clotting and haemorrhage, severe anaemia, or death in wounded fish. Haemorrhagic areas often appear in fragile tissues such as the gills.

Requirements

Nutritional deficiency signs in fish occur within 10-14 weeks. Quantitative requirement studies have not been completed are needed in order to define the vitamin K-requirement of rapidly growing young fish, especially those held in high population density under intensive fish husbandry conditions.

Sources and Protection

Vitamin K sources are green, leafy vegetables. Alfalfa leaves are among the best sources of vitamin K. Low levels are found in soybeans and animal liver. Synthetic menadione is also available. Although vitamin K in ground alfalfa is fairly stable, synthetic material should be protected from exposure to ultraviolet light and to excessive oxidizing or reducing conditions. The use of rapidly cured dehydrated alfalfa is essential to minimize formation of the physiological antagonist, dicumarol. The diet should be kept dry, prepared with minimum exposure to air oxidation, and fed as soon as practicable after manufacture to minimize vitamin K loss.

Antimetabolites and Inactivation

Dicumarol was isolated from spoiled sweet clover hay and shown to prevent the normal function of vitamin K in maintenance of normal blood clotting times. Early work was with cattle and hogs but was soon extended to rats, other experimental animals, and man. Dicumarol is an anticoagulant and has been used to prevent thrombosis in animals and man.

It is not an antimetabolite competing for vitamin K sites but plays another role in preventing normal blood clotting. Vitamin K counteracts the dicumarol effect quantitatively. Another anticoagulant, warfarin, is a common rat poison which has five to ten times the anticoagulant activity of dicumarol. The effects of warfarin can also be reversed by administration of vitamin K.

Clinical Assessment

Vitamin K status of fish is determined by measuring blood clotting time. Chemical methods for determining menadione content of active metabolite after storage are available but have not been applied to measure vitamin K activity in the liver or tissues of young, growing fish.

BIOTIN

Certain foods, notably liver and kidney, were found to be protective against a form of dermatitis resulting from the consumption of egg white by rats, in the early thirties. It was soon learned that a specific component of egg albumin, avidin, rendered dietary biotin unavailable, hence producing the symptoms. Biotin was variously called coenzyme R and 'vitamin H'. It was isolated by Du Vigneaud in 1941 and synthesized by workers of Merck and Company in 1943. Biotin was once thought to be part of factor H for fish. Blue slime patch disease due to biotin deficiency was reported in trout.

Chemistry

Biotin is a monocarboxylic acid slightly soluble in water and alcohol and insoluble in organic solvents. Salts of the acid are soluble in water. Aqueous solutions or the dry material are stable at 100 C and to light. The vitamin is destroyed by acids and alkalis and by oxidizing agents such as peroxides or permanganate. Biocytin is a bound form of biotin isolated from yeast, plant, and animal tissues. Other bound forms of the vitamin can generally be liberated by peptic digestion. Oxybiotin has partial vitamin activity but oxybiotin sulphonic acid and other analogues are antimetabolites inhibiting the growth of bacteria.

Avidin, a protein found in raw egg white, binds biotin and makes it unavailable to fish and other animals. Heating to denature the protein makes the bound biotin available again to the fish. Biocytin or ε-biotinyl lysine (the epsilon amino group of lysine and the carboxyl of biotin being combined in a peptide bond) is hydrolysed by the enzyme biotinase making the protein-bound biotin available.

Positive functions

Biotin is required in several specific carboxylation and decarboxylation reactions, including the carboxylation of pyruvic acid to form oxaloacetic acid. It is part of the coenzyme of several carboxylating enzymes fixing CO_2 such as propionyl coenzyme A involved in the conversion of propionic acid to succinic acid in methylmalonyl coenzyme A. Biotin is also involved in the conversion of acetyl CO_2 to malonyl coenzyme A in the formation of long chain fatty acids. It has possible involvement in citrulline synthesis and may have effects on purine and pyrimidine synthesis. It is involved in the conversion of unsaturated fatty acids to the stable cis form in the synthesis of biologically active fatty acids.

Deficiency Syndrome

Some signs of biotin deficiency in salmonids are skin disorders, muscle atrophy, lesions in the colon, loss of appetite, and spastic convulsions. Haematology discloses fragmentation of erythrocytes. Poor growth is a common symptom and has been reported for salmonids, common carp, goldfish (Carassius auratus) and eel. Blue slime patch disease in brook trout deficient in biotin appears typical for this species. Fish reared in 10 -15°C water exhaust biotin stores in 8-12 weeks and the first signs are anorexia, poor food conversion, and general listlessness before the more acute deficiency symptoms become detectable.

Requirements

The biotin requirements for young salmon, trout, carp, goldfish and eel raised under experimental conditions appear to be about the same. Brown trout appear to require nearly twice as much biotin in the diet as do brook or rainbow trout.

Sources and Protection

Rich sources of biotin are liver, kidney, yeast, milk products, and egg yolks. Nut meats contain good supplies of biotin. The diet should be protected from strong oxidizing agents or conditions which promote oxidation of ingredients. Raw egg white should not be incorporated into moist fish diets. Cooking will inactivate the avidin.

Antimetabolites and Inactivation

Raw egg white has already been discussed which irreversibly binds biotin and makes it unavailable to young fish. Many biotin homologues with different side chain lengths inhibit the growth of bacteria. Oxybiotin, in which sulphur is replaced by oxygen, has about the same biological activity as the natural biotin. On the other hand, oxybiotin sulphonic acid inhibits biotin activity.

Clinical Assessment

Measurement or urinary excretion of biotin in animals does not prove a good clinical method since biotin is synthesized by several organisms in the gut. In fish, this technique needs to be explored and may well be valuable as a clinical tool since the gastrointestinal tract of many freshwater fish contains only small quantities of bacteria. Biotin is one of the most expensive vitamins to add to fish rations. Salmon feeding actively in the oceans had liver biotin concentrations of 10-12 mg of wet liver tissue.

The concentrations in the liver of young salmon fingerlings in fresh water fed test diets containing an excess of biotin were between 6-8 mg of the vitamin/g of tissue. Fish with these levels of biotin in the liver should probably be in sound biotin nutritional status.

NIACIN

Nicotinic acid was synthesized in 1873 but was left on the shelf as an organic compound unrelated to the severe pellagra afflictions occurring throughout the world at that time. Sixty years later the compound was shown to be present in coenzymes I and II and two years thereafter Elvehjem cured 'black tongue' in dogs with the vitamin. Niacin was postulated to be part of factor H for fish in 1937, but deficiency symptoms were not adequately described until reported in trout in 1947.

Chemistry

Niacin is the preferred nomenclature. Nicotinic acid amide or niacinamide is the common form in which the vitamin is physiologically active. Niacin is a white, crystalline solid, soluble in water and alcohol. It is stable in the dry state and may be autoclaved for short periods without destruction. It is also stable to heat in mineral acids and alkali. Niacin is both a carboxylic acid and an amine and forms quaternary ammonium compounds because of its basic nature. Acidic characteristics include salt formation with alkali.

Niacin can be esterified easily, then converted to amides. Niacinamide is a crystalline powder soluble in water and ethanol and the dry material is stable up to about 60°C. In aqueous solutions it is stable for a short period when autoclaved. It is the form in which the vitamin is normally found in niacinamide adenine dinucleotide (NAD), and in niacinamide adenine dinucleotide phosphate (NADP).

Positive Functions

The major function of niacin in NAD and NADP is hydrogen transport in intermediary metabolism. Most of these enzyme systems function by alternating between the oxidized and reduced state of the coenzymes NAD-NADH and NADP-NADPH. Oxidation-reduction reactions may be anaerobic as when pyruvate acts as the hydrogen acceptor and lactate is formed, or the reactions may be coupled to electron transport systems with oxygen as the ultimate hydrogen acceptor; such aerobic reactions occur in respiration. Both NAD and NADP are involved in the synthesis of high energy phosphate bonds which furnish energy for certain steps in glycolysis, in pyruvate metabolism, amino acid and protein metabolism, and in photosynthesis. An interrelationship between thiamine and niacin exists since both vitamins are coenzymes in intermediary metabolism whereby food is oxidized to furnish energy for physiological functions, to maintain homeostasis, and for body temperature maintenance in homeotherms.

Deficiency Syndrome

Stores of niacin are more slowly exhausted under experimental conditions than are some of the other vitamins resulting in less defined and more slowly developing symptoms. Niacin deficiencies in fish were experimentally induced in the late forties and early fifties by using basal diets which had a low niacin content. Loss of appetite and poor food conversion were the first signs noted.

Then the fish turned dark and went off feed, followed by the appearance of lesions in the colon, erratic motion, oedema of the stomach and colon and muscle spasms while fish were apparently resting. A predisposition to sunburn in fish confined in the open in shallow ponds or raceways was described. Common carp showed a congestion of the skin with subcutaneous hemorrhages. Common symptoms of niacin deficiency in most fish studied were muscular weakness and spasms, coupled with poor growth and poor food conversion.

Requirements

Niacin requirements in young fish experimentally tested. In homeotherms on a balanced test ration the niacin requirement is generally estimated to be about ten times that of the thiamine requirement. These rations generally contain considerable carbohydrate material to furnish energy to maintain body temperature. In fish the requirement appears to be twenty to thirty times that of the thiamine needs determined for the same test conditions and test rations. This difference may be due to low carbohydrate content of young fish diets and the higher protein content of these rations. Conversion of tryptophan to niacin occurs in the mammalian liver and possibly also in the liver of fish. This conversion may account for the slow development of the niacin-deficiency syndrome in fish. However, after 10-14 weeks on diets devoid of niacin, deficiency symptoms did occur in several species of fish. The symptoms were reduced by replacement of niacin in the ration even when high protein diets containing an excess of tryptophan were fed. Too much niacin inhibits growth.

Sources and Protection

Niacin is found in most animal and plant tissues. Rich sources are yeast, liver, kidney, heart, legumes, and green vegetables. Wheat contains more niacin than corn and the vitamin is also found in milk and egg products. The vitamin is very stable since it is generally found in coenzyme form in raw materials. Niacin added to the diet as a supplement remains relatively unaltered during diet manufacture, processing, and storage.

Antimetabolites and Inactivation

Pyridine-3-sulphonic acid and 3-acetylpyridine are compounds structurally related to niacin and are antimetabolites for this vitamin in animals and in micro-organisms. Additional niacin can overcome the anti-metabolic effect. Deficiency symptoms in rats may be induced by 6-amino niacinamide. The symptoms are reversed by addition of ten times more niacinamide than the antimetabolite. Thioacetamide has been reported to be a niacin antagonist in fish.

Clinical Assessment

Salmon feeding actively in the oceans showed liver niacin content of 70-80μ g of niacin/g of fresh tissue. About half of this amount is present in fingerling salmon raised in fresh water at 12-15°C and fed test rations containing 40-50 per cent protein and niacin supplements 500-750 mg/kg dry diet. Urinary metabolities of niacin have been measured in other animals on standard niacin load in test rations containing a standard tryptophan load. The technique is well developed to measure the N^1-methyl derivative in mammalian urine.

These data have not been reported for fish but metabolism chambers are available for collecting branchial and urinary wastes from large fish intubated with different diet material.

FOLIC ACID (FOLACIN)

A factor effective in curing metaloblastic anaemia in monkeys was found in yeast and liver concentrates and was designated vitamin M in 1935. In 1939 an anti-anaemic factor was found in liver and called vitamin B_C. These were later shown to be the same substance with folic acid as the active ingredient. Folic acid was synthesized in 1946 and was soon used in fish diets for preventing purified diet anaemia.

Chemistry

Folic acid crystallizes into yellow spear-shaped leaflets which are soluble in water and dilute alcohol. it can be precipitated with heavy metal salts. It is stable to heat in neutral or alkaline solution, but unstable in acid solution. It deteriorates when exposed to sunlight, or during prolonged storage. Several analogues have biological activity including pteroic acid, rhizopterin, folinic acid, xanthopterin and several formyltetrahydropteroyl-glutamic acid derivatives.

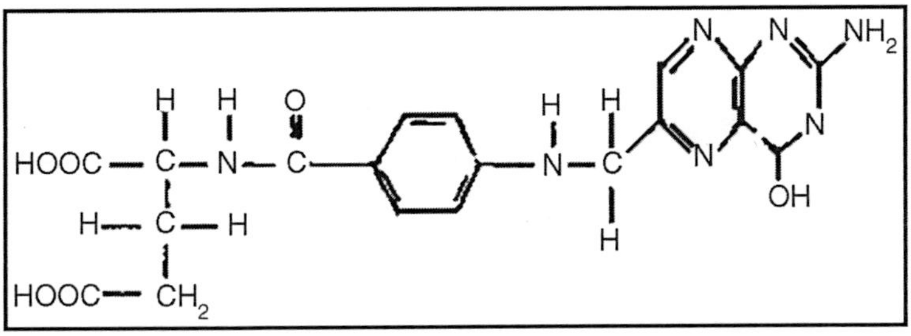

Fig. ***Folic Acid***

These have closely allied ring structures and many have been isolated as derivatives in various animals or microbiological preparations. One simple form, xanthopterin, present in the pigments of insects, is shown below and is of special interest because of early work with this compound as the anti-anaemic factor H for fish.

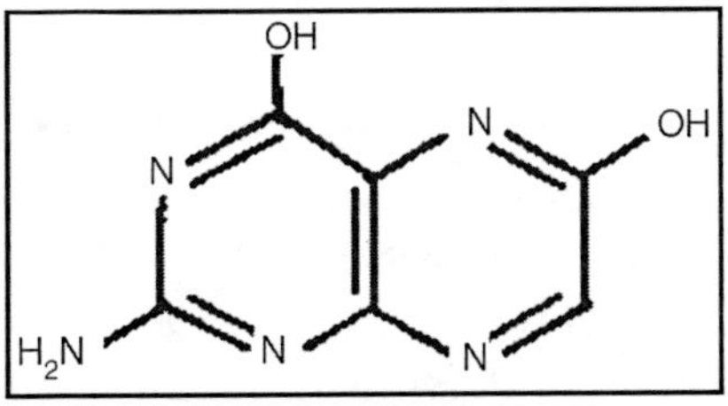

Fig. ***Xanthopterin***

Positive Functions

Folic acid is required for normal blood cell formation and is involved as a coenzyme in one-carbon transfer mechanisms. In the presence of ascorbic acid, folic acid is transformed into the active (5-formyl-5, 6,7,8) tetrahydrofolic acid. Folic acid is involved in many one-carbon metabolism systems such as serine and glycine interconversion, methionine-homocysteine synthesis, histidine synthesis, and pyrimidine synthesis.

Several coenzyme forms of the active vitamin have been isolated. Folic acid is involved in the conversion of megaloblastic bone marrow to normoblastic type. It has a role in blood glucose regulation and improves cell membrane function and hatchability of eggs.

Deficiency Syndrome

Macrocytic normochromic anaemia occurs in several experimental animals, including fish fed diets devoid of the folic acid. Increasing numbers of senile cells are observed as the deficiency progresses until only a few old and degenerating cells are found in the blood of deficient fish. Anterior kidney imprints disclose only adult cells and no preforms present. Other signs observed have been poor growth, anorexia, general anaemia, lethargy, fragile fins, dark skin pigmentation, and infarction of spleen.

Requirements

The requirement seems to be about the same for trout and salmon. Marginal macrocytic anaemias occur in fish fed diets containing marginal amounts of folacin. Individual fish ingesting adequate amounts of the vitamin show little variation in total erythrocyte counts.

Sources and Protection

Yeast, green vegetables, liver, kidney, glandular tissue, fish tissue, and fish viscera are good sources of folic acid. Insects contain xanthopterin which has folic acid activity. At one time the yellow pigment of xanthopterin was identified as the fish anti-anaemic factor H, but subsequent experiments showed only partial activity and that folic acid itself was a much more potent antimacrocytic anaemia factor. Insects may contribute significantly to the folic acid requirements of wild fish, but in scientific fish husbandry artificial diets are more reliable sources. Activity is lost during extended storage and when material is exposed to sunlight.

Therefore, dry feeds should be carefully protected during manufacture and moist diet rations should be carefully preserved. Both types of fish diets should be fed soon after manufacture to assure minimal loss of folic acid activity.

Antimetabolites and Inactivation

One antagonist of folacin is 4-aminopteroylglutamic acid or aminopterin.

This material, when incorporated in the diet of guinea pigs and rats, induces anaemia and leucopenia and has been used to treat leukemia in man. Amethopterin (4-amino-N^{10}-methylptero-ylglutamic acid) can also be used to induce deficiency by inhibiting purine, pyrimidine, and nucleic acid protection; *viz.*, inhibition of nucleic acid synthesis results in macroytic anaemia.

Clinical Assessment

Haematology is used as a simple clinical tool to assess haemopoiesis in fish.

Anterior kidney imprints easily disclose normal distribution of immature cells and preforms undergoing reticulosis.

Salmon actively feeding in the ocean and young salmon fingerlings raised in fresh water and fed diets rich in folacin show liver storage of 3-4μ g of folic acid/g of wet tissue. Microbiological assay is preferred for assessment of total folic acid in dietary raw materials because the total biological activity it measures includes all the various coenzyme forms and folic acid analogues.

Assessment of the dietary intake of folic acid is important for intensive cold water fish husbandry.

In pond culture, aquatic and terrestrial insects, algae, etc., may also be available and folic acid in the supplementary feed may not be as critical.

Since folic acid is labile in storage, excess amounts are generally added to manufactured feed in anticipation of storage losses. However, prudent fish husbandry dictates rapid use of manufactured rations with minimum storage. Routine periodic haematology of fish assures proper nutritional status for maximum production and sound health. It hasbeen noted in several series of experiments that when fish diseases occur through inadvertent contamination of the water supply, those groups of fish partially or completely deficient in folic acid were among the first to show acute disease symptoms. Therefore, folic acid must also play an important role in resistance to disease.

VITAMIN B_{12}

The antipernicious anaemia factor found to be contained in liver was isolated and crystallized in the mid forties. This substance, named vitamin B_{12} by its discoverers, was later to be recognized as essential for growth of chicken fed diets entirely of plant origin and was designated animal protein factor (APF). When anaemic salmon were injected with crystalline B_{12} in combination with folic acid and xanthopterin positive haemopoiesis occurred within a few days, and the salmon showed rapid recovery from the anaemia.

Chemistry

Vitamin B_{12} or cyanocobalamin has the following approximate chemical structure.

The molecule has a planar group and a nucleotide group lying nearly at right angles to one another. This cobalt-containing vitamin has a net charge of one at the central cobalt atom to which is attached a replaceable cyano group. Vitamin B_{12} is stable to mild heat in neutral solution, but is rapidly destroyed by heating in dilute acid or alkali. Crude concentrates are more unstable and rapidly lose activity. The compound is similar to the porphyrins in its spatial configuration with a central cobalt atom linked to four reduced pyrrole rings in the haeme series. Replacing the cyanide ion with a variety of anions produce derivatives which have comparable biological activities; *viz.*, hydroxocobalamin, nitritocobalamin, chlorocobalamin and sulphatocobalamin.

Positive Functions

Cyanocobalamin is involved with folic acid in haemopoiesis. It is required by many micro-organisms and is a growth factor for many animals. The animal protein factor present in fish and animal by-products was not recognized until crystalline vitamin B_{12} was injected into anaemic chinook salmon fingerlings in 1949 and positive haemopoiesis was observed. A coenzyme incorporating vitamin B_{12} is involved in the reversible isomerization of methylmalonyl coenzyme A to succinyl coenzyme A and in the isomerization of methylaspartate to glutomate.

Cyanocobalamin is involved in the coenzyme for the methylation of homocystine to form methionine. It is also involved in several other one-carbon reactions and in the synthesis of labile methyl compounds. One vitamin B_{12} containing coenzyme acts in methylation of the purine ring during thymine synthesis. Vitamin B12 is also involved in cholesterol metabolism, in purine and pyrimidine biosynthesis, and in the metabolism of glycols.

Deficiency Syndrome

Deficiency signs in young pigs, chicks, and rats show abnormal blood elements, poor growth, and pernicious anaemia. An intrinsic factor is necessary for good absorption of the vitamin from the gut. This factor is a low molecular weight mucoprotein which normally occurs in gastric juice, and especially in hog gut mucosa. Pernicious anaemia in chinook and coho salmon is characterized by fragmented, erythrocytes with many aberrant forms present. Haemoglobin levels are inconsistent and erythrocyte counts have a range extending from frank anaemia to a near normal blood pattern. Cyanocobalamin stores in fish tissues are slowly exhausted and only after 12-16 weeks on test do the symptoms appear in deficient salmon populations. Poor appetite, poor growth, poor food conversion, and some dark pigmentation can be observed before frank anaemia is detected.

Requirements

The estimates of requirement based on fish that showed normal haematology at the end of the 16-week feeding period, and maximum liver storage of the vitamin determined by Lactobaccilis leichmaniimicrobiological assay.

Sources and Protection

Rich sources of vitamin B_{12} are found in fish meal, fish viscera, liver, kidney, glandular tissues, and slaughter house wastes. Since vitamin B_{12} is labile on storage, and in mild acid solution is easily destroyed by heating, care must be exercised in diet preparation containing flesh or meat scraps.

Antimetabolites and Inactivation

The vitamin B_{12} coenzymes are very unstable under light, which rapidly decomposes the coenzymes. Photo-sensitivity is increased in dilute acid solutions.

Clinical Assessment

Generalized anaemia with fragmentation of erythrocytes and extremely variable haemoglobin levels and erythrocyte counts indicate possible B_{12} deficiency. Prompt response in individual fish is obtained by injecting B_{12} alone or in combination with folic acid in the ratio of 1 part vitamin B_{12} to 100 parts folic acid. Careful interpretation of haemotological data will enable one to distinguish one form of anaemia from the other.

Ascorbic Acid

Experimental work to cure scurvy with fruit juice was described by Lind in 1753, but nearly 200 years elapsed before the exact chemical compound responsible for reducing the symptoms was defined. The isolation of ascorbic acid was due to Szent-Györgyi of Hungary and C.G. King of the U.S.A. Vitamin C synthesis was accomplished in 1933 after the chemical structure of ascorbic acid was established by British and Swiss workers.

McCay and Tunison reported scoliosis in brook trout fed formalin-preserved meat in 1934 and McLaren observe haemorrhages in trout fed rations low in ascorbic acid. It was not until the sixties that a critical need for L-ascorbic acid by trout and salmon was demonstrated.

Chemistry

L-ascorbic acid is a white, odourless, crystalline compound, soluble in water but insoluble in fat solvents. It is readily oxidized to dehydroascorbic acid, the less biologically potent form. Ascorbic acid is very stable in acid solution because of the preservation of the lactone ring, but in alkaline solution hydrolysis occurs rapidly and vitamin activity is lost.

It is very heat labile and prone to atmospheric oxidation, especially in the presence of copper, iron, or several other metallic catalysts. The reduced form is the most biologically active but several derivatives or salts are obtainable which have varying degrees of ascorbate activity.

Positive Functions

L-ascorbic acid acts as a biological reducing agent in hydrogen transport. It is involved in many enzyme systems for hydroxylation; *i.e.*, hydroxylation of tryptophan, tyrosine, or proline. It is involved in the detoxification of aromatic drugs and also acts in the production of adrenal steroids.

Ascorbic acid is necessary for the formation of hydroxy proline which is a constituent of collagen, a component of intercellular material in bones and soft tissues. Ascorbic acid plays a synergistic role with vitamin E as intracellular antioxidants and free radical traps. The conversion of folic acid to folinic acid requires vitamin C. Ascorbic acid is involved in the formation of chondroitin sulphate and intercellular ground substance.

Labelled ascorbic acid fed to fish previously deficient in the vitamin was shown to be rapidly mobilized and fixed in areas of rapid collagen synthesis and became concentrated in the thick collagen of the skin and in cartilagenous bones, as well as in the glands of the anterior kidney. Ascorbic acid is also involved in erythrocyte maturation.

Deficiency Syndrome

Scurvy with impaired collagen formation, perifollicular haemorrhages, loose teeth and poor bone formation, anaemia, and oedema have been reported in other animals. Deficiency signs in fish are generally related to impaired collagen formation. Fish soon show hyperplasia of jaw and snout.

The same symptoms have been observed in trout, salmon, yellowtail, carp, guppies and char. Histologically, hypertrophy of the adrenal tissues and haemorrhage at the bases of fins have been observed in coho salmon.

Deficiency signs cease to develop and new growth becomes normal upon replacement of ascorbic acid in the ration. Anaemia eventually develops in extremely deficient fish and scoliosis and lordosis do not repair but are walled off by new growth around the afflicted areas of the spine when ascorbic acid is once again added to the ration.

Requirements

In the rainbow trout, reasonable blood and anterior kidney storage levels were obtained with an intake of about 100 mg of vitamin C/kg of dry ration in 10°, 12°, or 15° C water systems. When wound repair experiments were initiated, however, or when fish were exposed to other stress then the requirements doubled or tripled.

When severe abdominal or intramuscular wounds were inflicted, young fish needed at least 500 mg of active ascorbate for tissue repair comparable with control fish receiving 1 g or more of ascorbate in the diet/kg of dry diet.

Coho salmon appear to need about half of these requirements for adequate tissue levels and for maximum severe wound repair rates. The growth response and tissue repair for rainbow trout and coho salmon. The requirement for ascorbic acid is related to stress, growth rate and size of the animal, as well as to the other nutrients present in the diet.

A compromise value of about 200 mg of ascorbic acid/kg diet for trout and salmon raised in freshwater systems between 10-15°C would ensure reasonable tissue storage levels and furnish some excess for mild stress conditions and for ascorbic acid loss from the diet through oxidation.

Large common carp can synthesize some ascorbate and the requirement for this species may be dependent on fish size and the environment in which they are reared.

Table. Growth and Tissue Ascorbate

Vitamin C diet treatment	Trout			Salmon		
	Average weight at 24 weeks	Ascorbate concentrate[1/]		Average weight at 24 weeks	Ascorbate concentrate[1/]	
		blood	kidney		blood	kidney
mg/100 g	g	μg/g	μg/g	g	μg/g	μs/s
0	2.4	[2/]	[2/]	5.0	22.3±2.2	89
5	9.6	34.4±2.9	125	6.0	30.5±1.2	132
10	10.6	34.6±1.3	137	5.7	35.8±1.6	265
20	10.1	38.8±3.3	132	6.1	34.2±2.3	183
40	10.2	46.8±6.2	162	6.3	33.7±2.0	225
100	10.8	51.0±4.6	247	6.0	37.8±2.3	321

Note: [1/] Average of five samples for blood (± S.D.) and two for head kidney tissue

[2/] No fish available for assay

Sources and Protection

Ascorbic acid is widely distributed in nature with citrus fruits, cabbage, liver, and kidney tissue good sources for the vitamin. Fresh insects and fish tissues contain reasonable amounts of the vitamin. Synthetic ascorbic acid is also readily available. Fish food should be protected from oxidizing agents and kept sealed or frozen until used to prevent loss of the vitamin.

Antimetabolites and Inactivation

D-ascorbic acid, the optical isomer of the active form, has no vitamin activity and competes for sites of several enzyme reactions mediated by L-ascorbic acid. 6-deoxy-L-ascorbic acid, dehydroascorbic acid and L-glucoascorbic acid have very little activity.

Clinical Assessment

Ascorbic acid status for experimental animals is normally attempted by tissue ascorbate analysis. Most of the assays previously used measure total ascorbate and not biologically active L-ascorbic acid. Consequently these are fraught with errors and misconception of true vitamin C status. In fish tissues, blood and liver do not adequately reflect the ascorbic acid intake and status, whereas assay of the anterior kidney which contains adrenal tissue provides a fairly representative picture of tissue storage of the vitamin. Stress rapidly reduces the ascorbic acid content of this tissue with concurrent production of adrenal steroids. Conversely, dietary repletion is reflected by up to four or five-fold storage levels from the deficient state. Examination of fragile support cartilage in the gill filaments under low magnification will detect early hypovitaminosis before clinically acute symptoms become noticeable. However, the best tissue for routine clinical analysis to assess vitamin C status in trout and salmon appears to be the anterior kidney with samples selected from the junction of the two forward wings. Tissues are blotted free of blood with filter paper, and then assayed for total ascorbate by one of the improved quick methods to determine total ascorbate.

Inositol

Muscle 'sugar' was discovered by Scherer in 1850 and was characterized as inositol in 1887. Inositol was shown to be an alopecia (a form of hairlessness) preventing factor for mice in 1940. Poor growth and poor food passage in inositol-deficient fish was observed in salmon and carp.

Chemistry

Seven optically inactive and two optically active isomers of hexahydroxycyclohexane can exist. One of the optically active forms, myo-inositol, is a white crystalline powder soluble in water and insoluble in alcohol and ether.

The material can be synthesized, but is easily isolated from biological material in free or combined forms. The mixed calcium-magnesium salt of the hexophosphate is phytin. Isomers have little biological activity but will compete in chemical reactions. Inositol is a highly stable compound.

Positive Functions

Myo-inositol is a structural component in living tissues. It has lipotropic action by preventing accumulation of cholesterol in one type of fatty liver disease and is involved with choline in maintaining normal lipid/metabolism. It is a growth-promoting substance for micro-organisms and prevents an alopecia in mice. It is a reserve carbohydrate in muscle as well as a major component of phosphoglycerides in animal tissues.

Deficiency Syndrome

Poor growth, increased gastric emptying time, oedema, dark colour, and distended stomachs are symptoms observed in salmon, trout, carp and catfish held for long periods on inositol-deficient test rations. A 'spectacle-eye' condition described for rats has not been observed under the experimental conditions used in fish studies. The major deficiency sign is inefficiency in digestion and food utilization and concomitant poor growth leading to a population of fish with distended abdomens.

Requirements

Inositol needs have been measured in only two species of salmon and carp and the requirement is high for maximum growth and for maximum liver storage. Whether liver storage is accurate criteria to determine requirements is debatable since inositol intake was compared with maximum growth rate and diet conversion to develop a tentative requirement for young fish for this 'muscle' sugar.

Sources and Protection

Myo-inositol occurs ubiquitously in large amounts wherever biological tissue is found. Wheat germ, dried peas, and beans are rich sources. Brain, heart, and glandular tissues are very good sources of biologically active inositol. Citrus fruit pulp and dried yeast also contain inositol. The compound is stable.

Antimetabolites and Inactivation

Seven optically inactive and one optically active but biologically inactive stereo isomers occur. Since inositol is synthesized in the biologically active form by many microorganisms in the gut, only large quantities of chemically synthesized inactive isomers added to diets would interfere with the metabolism of inositol for growth and normal physiological function.

Biologically inactive stereoisomers of myo-inositol do not compete for critical sites in metabolism. Methyl derivatives and mono-, di-, and triphosphoric acid esters occur naturally. Salts of the hexaphosphate or phytin make the bound inositol practically unavailable to the animal.

Clinical Assessment

Assessment in fish has been based on lack of deficiency signs coupled with the most efficient food conversion. Salmon feeding actively in oceans show 1-1.5 mg of inositol/g of fresh liver tissue and young fingerlings raised in freshwater at 10-15°C had 600-700μ g/g of liver tissue. An alternate, better assessment may be based on a standard muscle or whole carcass analysis for free or for bound inositol. Projection of inositol intake from normal fish diet ingredients should indicate an excess of this particular vitamin.

Choline

Methylation as a basic metabolic process was postulated by Hoffmeister. Methyl transfer was shown in vivo by Thompson and the interrelationships between choline, methionine, and homo-cystine were shown by du Vigneaud in 1939-42. Trout fed low choline rations developed haemorrhagic kidneys and salmon showed an aversion to food in choline-deficient diet.

Chemistry

Choline is a very strong organic base and forms many derivatives widely distributed in animal and vegetable tissue. One derivative, acetylcholine, is involved in the transmission of nerve impulses across synapses. Choline is very hygroscopic, very soluble in water, and is stable to heat in acid, but decomposes in alkaline solutions.

Positive Functions

Choline acts as a methyl donor in trans-methylation reactions. It is a lipotropic and antihaemorrhagic factor preventing the development of fatty livers. It is involved in the synthesis of phospholipids and in fat transport.

Acetylcholine transmits the excitory state across the ganglionic synapses and neuromuscular junctions. Choline is essential for growth and good food conversion in fish.

Deficiency Syndrome

Deficiency signs include poor growth and poor food conversion accompanied by impaired fat metabolism. Haemorrhagic kidneys and intestines have been reported in trout and increased gastric emptying time has been observed in salmon.

Sources and Protection

Rich sources of choline are wheat germ, beans, brain, and heart tissue. Choline hydrochloride, the commercially available form, may inactivate α - tocopherol and vitamin K when in direct contact with these vitamins and care should be exercised in selecting properly protected (gelatin coated) fat soluble vitamins in fish diet preparation.

Clinical Assessment

Choline status of fish can be estimated from the assay of choline content of the dietary ingredients with the absence of deficiency signs. Maximum liver storage may not be the best criteria to determine choline nutritional status but has been used to assess the tentative requirement listed for the two species of salmon.

p-Aminobenzoic Acid

Para-aminobenzoic acid is a white crystalline powder which is water soluble and heat and light stable in aqueous and mild alkaline solution. It is a growth promoting vitamin for micro-organisms which require the vitamin for folic acid synthesis. Large intake has been shown to counteract the antimetabolite effect of sulphonamides in bacterial culture. No positive function or deficiency signs have been observed in fish.

Lipoic Acid

Lipoic acid is both fat soluble and water soluble. Its active form is the amide derivative known as lipoamide. Lipoic acid functions as a coenzyme in a -ketoacid decarboxylation. It was discovered independently in several laboratories during the period 1945-50 and shown to be an essential component of multienzyme, the pyruvate dehydrogenase complex.

Pyruvate is converted to 'active acetaldehyde' which in turn is picked up by lipoamide. Subsequent oxidation of the aldehyde results in the reduction of lipoamide to a disulphydryl form.

The multi-enzyme unit also includes thiamine pyrophosphate, coenzyme A, and flavin adenine dinucleotide. Glandular tissues are good sources of lipoic acid. No requirements have been determined for fish.

WATER-SOLUBLE VITAMINS

Water-soluble vitamins include eight well-recognized members of the vitamin B complex: thiamine, riboflavin, pyridoxine, pantothenic acid, niacin, biotin, folic acid and vitamin B12; the water-soluble essential nutritional factors: choline, inositol, ascorbic acid; and vitamins with less-defined activity for fish: p-aminobenzoic acid, lipoic acid and citrin.

The first eight, though required only in small amounts in the diet, play major roles in growth, physiology and metabolism. Choline, inositol and ascorbic acid are required in appreciable quantities in the diet and sometimes are not referred to as vitamins but as major dietary nutrients.

Thiamine

Experimental avian polyneuritis, a condition resembling beriberi, was produced by Eijkman in Java in 1886 and the anti-beriberi factor was crystallized and named "vitamine". The term "thiamine" was introduced when the chemical nature of the factor was established. Thiamine was isolated from rice polishings in 1926 and synthesized in 1936.

Chemistry

Thiamine hydrochloride is a water-soluble, colourless, monoclinic, crystalline compound. It is comparatively stable to dry heat but is rapidly broken down in neutral or alkaline solutions and is split by sulphites into constituent pyrimidine and thiazole moieties. It has a characteristic yeast-like odour.

The pyrimide ring is relatively stable, but the thiazole ring is easily opened by hydrolysis. Several derivatives are stable to heat and appear to be more completely soluble in weak alkaline solutions than thiamine itself and still show biological activity in animals. These derivatives include thiamine propyl disulphide, benzoylthiamine disulphide, dibenzoylthiamine, and benzoylthiamine monophosphate. Both thiamine hydrochloride and thiamine mononitrate have been successfully used as the active vitamin in test diets for fish nutrition studies.

Positive Functions

Thiamine functions in all cells as the coenzyme cocarboxylase, thiamine pyrophosphate, which participates in the oxidative decarboxylation of pyruvic acid to acetate for entry into the tricarboxylic acid (TCA) cycle.

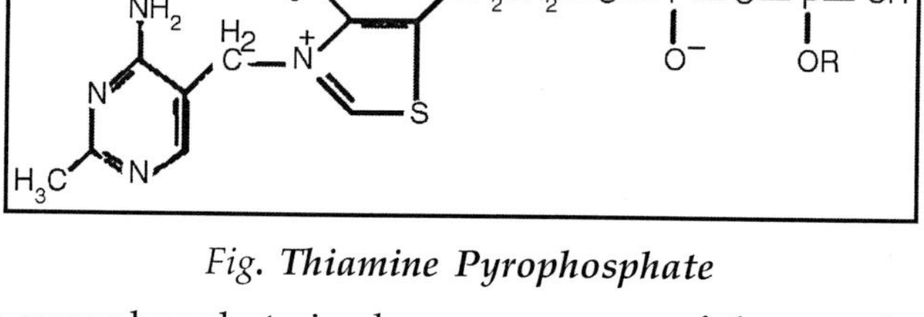

Fig. Thiamine Pyrophosphate

Thiamine pyrophosphate is also a coenzyme of the transketolase system by which direct oxidation of glucose occurs in the cytoplasm of cells via the pentose phosphate pathway. Erythrocyte (RBC) levels of metabolites of this system have been used to indicate thiamine status in experimental animals, including salmon and trout.

Thiamine is essential for good appetite, normal digestion, growth and fertility. It is needed for normal functioning of the nervous tissue and the requirement is determined by the caloric density of the diet.

Deficiency Syndrome

Deficiency signs in salmonids include impaired carbohydrate metabolism, nervous disorders, poor appetite, poor growth, and increased sensitivity to shock. A trunk-winding symptom in eels has been reported, together with haemorrhage at the base of the fins. Skin congestion and subcutaneous haemorrhage occurs in carp fed thiamine-deficient diets.

Typical symptoms observed in salmonids, carp and catfish. Thiamine deficiency has also been reported in marine flatfish started on clam neck diets stored long enough for thiaminase present to hydrolyse the thiamine in the ration. Typical nervous paralysis occurred with rapid mortality from physical shock.

Requirements

In determining dietary thiamine requirements of fish, some considerations must be placed upon the composition of dietary ingredients in the ration. Fat content of the diet may affect not only caloric intake but also the thiamine requirement because cocarboxylase participates in the oxidation of that through a-ketoglutarate. Therefore, fish on a high fat diet and low thiamine intake might take longer to develop deficiencies and will give an erroneous requirement. The thiamine requirements of selected fish species.

Sources and Protection

Common sources for thiamine are dried peas, beans, cereal bran, and dried yeast. Fresh glandular tissue is also a good source for thiamine and other members of the vitamin B water-soluble complex, but is seldom used in modern commercial fish diets. Thiamine can be easily lost by holding diet ingredients too long in storage or by preparing the diet under slightly alkaline conditions or in the presence of sulphide.

Wet or frozen diets pose a problem because moisture content increases the chance of enzymatic (thiaminase) hydrolysis and subsequent destruction of thiamine. Obviously, wet or moist diet, preparations containing any fresh fish or shellfish tissue must be used immediately.

Antimetabolites and Inactivation

Acetylcholine is an antagonist to thiamine and pyrithiamine. Oxythiamine and normal butylthiamine are specific antimetabolites. Several thiaminases occur which destroy thiamine. These rupture the thiazole ring at the sulphur bond making the residue inactive. Freshwater fish tissues have high thiaminase activity as do tissues from clams, shrimp and mussels.

Thiaminases have also been found in beans and mustard seed and in several micro-organisms.

Thiaminase activity is low in most saltwater fish tissues, however, and the enzyme is inactivated by heating or prolonged pasteurization. Thiamine present in fresh Torula yeast is relatively unavailable to fish, but the yeast becomes an excellent thiamine source after rupturing the cells by steam treatment or by dehydration.

Clinical Assessment

Clinical assessment of thiamine status have been made by measuring erythrocyte transketolase activity. Typical saturated levels for thiamine activity in sea salmon range from 15 to 20 g of thiamine/g of wet liver tissue. Fingerling chinook or coho salmon reared in 10 or 15°C water on test diets containing the listed amount or more of thiamine hydrochloride: assayed at 8-10 g of thiamine/g of wet liver tissue. These liver storage levels and normal erythrocyte transketolase activity in the absence of any deficiency sign will indicate an adequate thiamine intake for that fish population.

PYRIDOXINE

A new factor which could cure dermatitis in rats was reported by Györgi in 1935 and was named vitamin B_6. The name pyridoxine is also attributed to Györgi.

The active material was isolated in 1939 and pyridoxine was subsequently synthesized. First quantitative requirements for fish were described in 1944. Pyridoxine deficiency has since been reported in trout and in salmon.

Positive Functions

Pyridoxal phosphate is the coenzyme, codecarboxylase, involved in the decarboxylation of amino acids. It is also the co-factor of the 22 different transaminases present in animal tissues. As codecarboxylase, pyridoxal phosphate has been shown to participate in the decarboxylation of 5-hydroxytryptophan to produce 5-hydroxytry-ptamine or serotonin. As a co-factor of the enzyme desulphydrase converts cysteine to pyruvic acid. Pyridoxal phosphate is the co-factor for the synthesis of δ-amino-levulinate, the latter a precursor of haeme. Many neuro hormones require pyridoxal phosphate as a coenzyme in their synthesis.

Pyridoxine is also involved in fat metabolism, especially of the essential fatty acids. It is involved in the synthesis of messenger RNA which determines amino acids sequence in polypeptide synthesis. Pyridoxine plays a most important role in protein metabolism and, as a result, carnivorous fish have stringent requirements for the vitamin in the diet and stores are rapidly exhausted.

Deficiency Syndrome

Signs of pyridoxine deficiency in fish. Since salmonids, ictalurids, and very young cyprinids are carnivorous, with protein requirements for young animals between 40-50 per cent of the ration, pyridoxine stores are rapidly exhausted when fish are held on pyridoxine-deficient rations. Acute deficiency signs occur in salmon after 14-21 days on a high protein diet of pyridoxine and the entire population dies within 28 days in 12-15°C water.

Deficiency is followed by general nervous disorders, and alteration in control of melanophore contraction. Postmortem rigor mortisoccurs very rapidly. Rapid and gasping breathing with flexing of the opercules is a common observation and oedema in the peritoneal cavity with colourless serous fluid occurs in fish on some experimental treatments. Salmon, trout, carp and yellowtail exhibit premortem rigor a few hours before death. Recovery at late stage deficiency is very unlikely unless fish are injected with pyridoxal phosphate. Handling the animals generally induces more damage than vitamin administration will correct. Recovery among fish still feeding is equally rapid and dramatic upon administration of pyridoxine hydrochloride in the diet. Deficiency signs disappear within a day or two. Erythrocyte and plasma transaminase activities reflect the deficiency state.

Requirements

Pyridoxine requirements for trout, salmon, carp, and yellowtail. Supplementing commercial fish diets with pyridoxine is desirable because of the important role the vitamin plays in protein metabolism.

Sources and Protection

Good sources of pyridoxine are yeast, whole cereals, egg yolk, liver and glandular tissues. Pyridoxine compounds in phosphorylated form present in agricultural products are fairly stable but are sensitive to ultraviolet radiation. Some pyridoxal phosphate will be lost on exposure to air. Free forms or pyridoxal and pyridoxamine are rapidly destroyed by air, light and heat when moist.

Antimetabolites and Inactivation

Antagonists may compete for reaction sites of the apoenzyme or may react with pyridoxal phosphate to form inactive compounds. Deoxypyridoxine is a potent pyridoxine antagonist because of competition for apoenzyme sites but is a useful agent to accelerate pyridoxine deficiency in experimental animals. This same compound inhibits tyrosine decarboxylase. Methoxypyridoxine is another antagonist. Oxopyrimidine (2-methyl-4-amino-5-hydroxymethyl-pyrimidine) produces liver damage and inhibits glutamate decarboxylase.

Clinical Assessment

Plasma and erythrocyte transaminase activity in most instances reflects pyridoxine status of fish although high tryptophan load in the diet tends to increase the pyridoxine requirement. Liver storage measured by microbiological assay showed 5-6 V-g of pyridoxine activity/g in fresh sea salmon liver; whereas, fingerling salmon fed a 50 per cent protein diet in freshwater had 2-3 mg/g of wet tissue. Assay of diet for pyridoxine by microbiological methods gives a truer representation of total pyridoxine activity including intermediates.

RIBOFLAVIN

Following the recognition that the original water-soluble B consisted of more than one factor, a new growth-promoting factor consisting of yellow-green pigments was isolated in 1879. This new factor was designated G by some and 83 by others. The second water-soluble vitamin discovered was given its specific chemical name riboflavin.

Lactoflavin, hepatoflavin, and ovoflavin were also shown to be identical with the pure riboflavin. Riboflavin occurs in the free form only in the eye, whey and urine.

Chemistry

Riboflavin is a yellow-brown crystalline pigment. The vitamin is very slightly soluble in water but soluble in alkali. It is insoluble in most organic solvents. Riboflavin is stable to oxidizing agents in strong mineral acids and in neutral aqueous solution. It is also stable to dry heat but is irreversibly decomposed on irradiation with ultraviolet rays or visible light, breaking down to lumiflavin. Riboflavin phosphate is the chemically active group of Warburg's 'yellow' enzyme.

Positive functions

Riboflavin functions in the tissues in the form of flavin adenine dinucleotide (FAD) or as flavin mononucleotide (FMN). The flavo-proteins function as enzymes of tissue respiration and are involved in hydrogen transport to catalyse the oxidation of reduced pyridine nucleo-tides (NADH and NADPH). Thus, they function as coenzymes for many oxidases and redunctases such as cytochrome e reductase, D- and L-amino acid oxidases, xanthine and aldehyde oxidase, succinic dehydrogenase, glucose oxidase and fumaric dehydrogenase. Riboflavin is also involved with pyridoxine in the conversion of tryptophan to nicotinic acid and is most important in the respiration of poorly vascularized tissues such as the cornea of the eye. Riboflavin is involved in the retinal pigment during light adaptation and lack of it causes impaired vision and photophobia in experimental animals, including fish.

Deficiency Syndrome

Tissue riboflavin is exhausted in young salmonids after 10 to 12 weeks on riboflavin-deficient diets in 10-15 °C water systems. Poor appetite and poor diet efficiency are the first signs, followed by photophobia, mono or bilateral cataract's, corneal vascularization, eye haemorrhage, incoordination, and general anaemia. Dark pigmentation, coupled with striated constrictions of abdominal wall in salmon have been noted. Skin atrophy has been reported for some fish species and abnormal pigmentation of both skin and iris has been noted. Replacement of riboflavin in the diet reduced the symptoms except when cataracts have developed.

This irreversible condition will continue in monolateral cataracts throughout the life of the fish, whereas bilateral cataracts largely result in eventual starvation and death of the afflicted animal.

The first specific signs have consistently appeared in and about the eyes of salmonids, carps and catfish.

Requirements

The requirements of fish for riboflavin under experimental conditions in 10-15 °C water supplies. Values for the trout are slightly lower than those reported for salmon. The requirements may vary depending upon the balance of other dietary ingredients, caloric density and environmental conditions under which the fish is raised.

The values established under these standard test conditions should serve as tentative-levels which will satisfy biological demands for normal growth, health, and physiological function. Most of these studies have been made on very young fish, often initially feeding fry, with the logical assumption that their vitamin needs would be more than that of larger fish having advanced metabolic enzyme systems capable of synthesizing at least some of these vitamins to partially meet the requirements.

Sources and Protection

Riboflavin is widely distributed in plants and in animal glandular tissues. Milk, liver, kidney, heart, yeast, germinated grains, peanuts, soybeans and eggs are rich sources. Keeping feeds from sunlight or intense artificial light is necessary to minimize loss of the vitamin by conversion to lumiflavin.

Antimetabolites and Inactivation

When the ribose group in the molecule is replaced by other groups, analogues are formed which either have reduced activity or become antimetabolites. Galactoflavin is an antagonist to riboflavin and inhibits growth of rats when the diet contains this compound. Flavin monosulphate inhibits D-amino acid oxidase and appears to act as a competitor and inhibitor for growth of Lactobacillus casei. The hydroxyethyl analogue is an antagonist for riboflavin function in both rats and bacteria and also shows antifungal activity.

Clinical Assessment

Liver tissue of actively feeding sea salmon contains 6-8 g of riboflavin/ g of wet tissue. In a freshwater environment, young fish fed test diets in 15 ºC water systems showed liver storage of 3.5-4.0 μg/g. Riboflavin content of blood plasma does not change significantly in riboflavin deficiency in experimental animals. However, the erythrocyte riboflavin content has been reported at around 10μ g/100 ml blood for man on low riboflavin intake and approximately twice that level on high intake. Urinary excretion of riboflavin has been used clinically to indicate dietary status, with 200 μg/24 it suggesting adequate intake and less than 100μ g indicating low intake. Excretion of 50μ g or less daily is strong indication of extended dietary deficiency. Several studies show that excretion of less than 200μ g of riboflavin/g of creatinine would be indicative of deficiency.

PANTOTHENIC ACID

A chick dermatitis was cured by Elvehjem and Köehn in 1935 using a factor containing β-alanine. The term pantothenic acid was given earlier by Williams to a yeast growth factor which was later recognized as the anti-dermatitis factor. Pantothenic acid was synthesized by Stiller in 1940. Phillips observed clubbed gills in trout fed pantothenic acid-deficient diets in 1945. Rucker detected the same condition in salmon fed low pantothenic acid diets.

Chemistry

Pantothenic acid may be considered as dihydroxydimethylbutyric acid bonded to β-alanine. The free acid is a yellow, viscous oil and therefore the compound generally used in fish diet preparation is the calcium salt. This salt is a white crystalline powder readily soluble in water, mild acid, and is almost insoluble in organic solvents. It is stable to oxidizing and reducing agents and to autoclaving, but is labile to dry heat, hot alkali, or hot acid. Pantothenol has almost as much activity as pantothenic acid for growth of chicks. Pantothenic acid acetate, benzoate, and diphosphate esters are biologically active for animals but not for lactic acid bacteria. The optical isomer L-pantothenic acid, appears physiologically inert. Some organisms may utilize a portion of the molecule. Bacteria appeared to require only the dihydroxydimethylbutyric acid, and some yeasts utilize only g-alanine. Animals, however, need the entire pantothenic acid molecule or its reduced alcohol form.

Positive Functions

Pantothenic acid is part of acetyl coenzyme A which occurs in many enzymatic processes involving 2-carbon compounds. It has been shown to be required by all animal species studied and by many micro-organisms. The acetyl coenzyme A system is involved in the acetylation of aromatic amines and choline; condensation reactions for synthesis of acetate, fatty acids, and citrate; the oxidation of pyruvate and acetaldehyde; and is essential for the development of the central nervous system. The 2-carbon fragment called 'active acetate', or acetyl coenzyme A, is an essential intermediate in metabolism. It is involved in acylation of acetate, succinate, benzoate, propionate and butyrate. Pantothenic acid is involved in adrenal function and for the production of cholesterol. Coenzyme A is also involved in many other steps of intermediate metabolism of carbohydrates, fats and proteins.

Deficiency Syndromes

Under standard test conditions with deficient diet fed in 10-15°C water systems, salmon and trout exhaust pantothenic acid stores within 8-12 weeks. Fish stop feeding and close examination of gill filament show proliferation of epithelial surface plus swelling and clubbing together of the filaments and lamellae. The opercules become distended and the surface of the gills is often covered with an exudate. Fish become prostrate or sluggish. Necrosis, scarring and cellular atrophy of the tender gill elements occur and anaemia develops after prolonged deficiency. Dietary gill disease has been adequately described and correlated with pantothenic acid deficiency. The same type of symptom has been observed in salmon, trout, eel, carp and catfish. After replacement of pantothenic acid in the diet, recovery is rapid for those fish still feeding and gross deficiency symptoms disappear after about four weeks on the recovery diet although evidence of necrosis and scarring remains.

Sources and Protection

Good sources for pantothenic acid are cereal bran, yeast, liver, kidney, heart, spleen and lung. Fish flesh is a relatively rich source, although the content is only about 20 per cent of pantothenic acid found in animal glandular tissue.

Royal jelly probably contains the greatest amount as it contains over 500μ g of pantothenol/g dry weight. The calcium or sodium salt of pantothenic acid is relatively stable and can be incorporated into either moist or dry fish diets. Some loss is incurred during autoclaving and excessive heat should therefore be minimized during diet preparations. Since the free acid is labile to heat and also to acid and alkali, some loss can be expected during moist diet preparation or during storage. Certain cereal brans may have pantothenic acid bound in a form unavailable to fish because of the low digestibility coefficient and should not be relied upon as the sole pantothenic acid source in the diet.

Antimetabolites and Inactivation

Since pantothenic acid affects the respiration of many types of cells, compounds like 6-mercaptopurine, 2,6-diaminopurine, and 8-azaguanine, which inhibit growth of tumors are antagonistic towards pantothenic acid. Pantoyltaurine is an antimetabolite of pantothenic acid and has been used to accelerate deficiency syndromes in experimental animals.

Methyl-ω -pantothenic acid was reported to interfere with the formation of acetyl coenzyme A and accelerates deficiency syndromes in animals. This compound inhibits sulfanilamide acetylation in pigeon liver homogenates but did not prevent citric acid formation. Pantothenic acid can be used to overcome these inhibitory effects, including the reversal of mitosis blockage by 7-mecaptopurine in animals. High levels of calcium pantothenate in the diet also has a transient effect on the migratory urge of salmon.

Clinical assessment

Assay of pantothenic acid content in the diet may be misleading unless care is exercised in proper hydrolysis of the raw materials being assayed since pantothenic acid is only slowly liberated by normal hydrolytic procedures. Complete hydrolysis with enzyme preparations will liberate all of pantothenic acid from biologically active material of glandular tissues, fish flesh, yeast, and bran.

Salmon feeding actively in the oceans showed liver pantothenic acid content of 18-20μ/g of fresh tissue. Young chinook and coho salmon fingerlings reared in fresh water at 12-15°C showed maximum liver storage at about 14-16μ g of pantothenic acid/g of fresh liver tissue. Load tests and acetylation reactions of sulfanilamide have been used in man and other experimental animals but have not yet been extended for assessment of pantothenic acid status in fish.

5

Composition of Balanced Diet

GOOD DIETS

Low Fat-30 Diet: It is a weight reduction program for all who want to reduce their weight together with a group. Participation is for the time being limited to Germany. The most important facts about low fat nutrition is given in small groups. Meals should not contain more than 30% of calories coming from fat. You should eat only when feeling hungry. At the first sign of satiation one should finish the meal. The diet is based on points. Minimum amounts of foods are given. Once a week brochures on special topics are given. A long-lasting modification of the nutritional habits are essential in order to avoid overweigh.

Natural Weight Program: This program was developed by Dr. Doris Wolf She herself had been overweighted and tried to amend it using here knowledge as a psychologist and psychotherapist developing the Natural Weight Program.

Putting on weight because of emotional problems and stress are the main reasons for many people to skip all good intentions of an healthy nutrition. Frustration, anger and other negative emotions are compensated by eating. Alternatives to get emotionally balanced other than overeating are strategies of modern psychology. How to to handle one's emotions and how to accept one's own body not as an enemy, but as a fried, are important subjects of the Program. The negative attitude towards their body must be changed.

According to Dr.Doris Wolf all kind of diets are unsuitable for the majority of persons. There are biological and psychological reasons which are responsible for the fact that only 2 to 5 per cent of those who follow a diet register lasting weight reduction.

Dr. Wolf uses four principles as basis for her Program: The four principles:

1. Eat only when you are physically hungry and not when mental hungry.
2. Eat what you like for the moment.

3. Eat consciously, being aware of what you eat and how your body reacts.
4. Stop eating when you feel satisfied

VITAMINS

The term vitamin was coined from the words 'vital amine' as early scientists felt these chemicals which are vital for life were amines. Vitamins were discovered one at a time from 1900 to 1950, some as a cure for classic diseases such as beri-beri, pellagra, and scurvy, while others were discovered after research on various body functions. Vitamins like carbohydrates, proteins, and fats are organic compounds.

Unlike these nutrients, vitamins are required in minute quantities and are also called micronutrients. They do not provide energy and are present in very small quanti-ties in food, but nonetheless are vital for life processes. All vitamins can be synthesized on a commercial scale, though fresh foods are always preferred.

Definition Vitamins is the term used for a group of potent organic compounds other than proteins, carbohydrates, and fats which occur in minute quantities in food and which are essential for some specific body functions such as regulation, maintenance, growth, and protection. Many of them cannot be synthesized, at least in adequate amounts, by the body and must be obtained from the diet.

Classification

Vitamins are grouped according to their solubility in either fat or water. Fat-soluble vitamins. The fat-soluble vitamins are vitamins A, D, E, and K. They require fat for their absorption and can be stored in the body. If their intake is poor, but body stores are ample, deficiency symptoms will not be seen immediately.

Water-soluble vitamins. The water-soluble vitamins are B-complex vitamins and vitamin C. Being water soluble they are easily absorbed and the excess consumed is excreted in the urine. They are not stored in the body.

VITAMIN A

Vitamin A (or Vitamin A Retinol, retinal, and four carotenoids including beta carotene) is a vitamin that is needed by the retina of the eye in the form of a specific metabolite, the light-absorbing molecule retinal, that is necessary for both low-light (scotopic vision) and color vision. Vitamin A also functions in a very different role, as an irreversibly oxidized form of retinol known as retinoic acid, which is an important hormone-like growth factor for epithelial and other cells. In foods of animal origin, the major form of vitamin A is an ester, primarily retinyl palmitate, which is converted to the retinol (chemically an alcohol) in the small intestine.

The retinol form functions as a storage form of the vitamin, and can be converted to and from its visually active aldehyde form, retinal. The associated acid (retinoic acid), a metabolite that can be irreversibly synthesized from vitamin A, has only partial vitamin A activity, and does not function in the retina for the visual cycle. All forms of vitamin A have a beta-ionone ring to which an isoprenoid chain is attached, called a *retinyl group*. Both structural features are essential for vitamin activity.

The orange pigment of carrots – beta-carotene – can be represented as two connected retinyl groups, which are used in the body to contribute to vitamin A levels. Alpha-carotene and gamma-carotene also have a single retinyl group, which give them some vitamin activity. None of the other carotenes have vitamin activity. The carotenoid beta-cryptoxanthin possesses an ionone group and has vitamin activity in humans.

Vitamin A can be found in two principal forms in foods:

- Retinol, the form of vitamin A absorbed when eating animal food sources, is a yellow, fat-soluble substance. Since the pure alcohol form is unstable, the vitamin is found in tissues in a form of retinyl ester. It is also commercially produced and administered as esters such as retinyl acetate or palmitate.
- The carotenes alpha-carotene, beta-carotene, gamma-carotene; and the xanthophyll beta-cryptoxanthin (all of which contain beta-ionone rings), but no other carotenoids, function as vitamin A in herbivores and omnivore animals, which possess the enzyme required to convert these compounds to retinal. In general, carnivores are poor converters of ionine-containing carotenoids, and pure carnivores such as cats and ferrets lack beta-carotene 15,15'-monooxygenase and cannot convert any carotenoids to retinal (resulting in *none* of the carotenoids being forms of vitamin A for these species).

History

The discovery of vitamin A may have stemmed from research dating back to 1906, indicating that factors other than carbohydrates, proteins, and fats were necessary to keep cattle healthy. By 1917 one of these substances was independently discovered by Elmer McCollum at the University of Wisconsin–Madison, and Lafayette Mendel and Thomas Burr Osborne at Yale University. Since "water-soluble factor B" (vitamin B) had recently been discovered, the researchers chose the name "fat-soluble factor A" (vitamin A). In 1919, Steenbock (University of Wisconsin) proposed a relationship between yellow plant pigments (beta-carotene) and vitamin A. Vitamin A was first synthesized in 1947 by two Dutch chemists, David Adriaan van Dorp and Jozef Ferdinand Arens.

Equivalencies of Retinoids and Carotenoids (IU)

As some carotenoids can be converted into vitamin A, attempts have been made to determine how much of them in the diet is equivalent to a particular amount of retinol, so that comparisons can be made of the benefit of different foods.

The situation can be confusing because the accepted equivalences have changed. For many years, a system of equivalencies in which an international unit (IU) was equal to 0.3 μg of retinol, 0.6 μg of β-carotene, or 1.2 μg of other provitamin-A carotenoids was used. Later, a unit called retinol equivalent (RE) was introduced. Prior to 2001, one RE corresponded to 1 μg retinol, 2 μg β-carotene dissolved in oil (it is only partly dissolved in most supplement pills, due to very poor solubility in any medium), 6 μg β-carotene in normal food (because it is not absorbed as well as when in oils), and 12 μg of either α-carotene, γ-carotene, or β-cryptoxanthin in food.

Newer research has shown that the absorption of provitamin-A carotenoids is only half as much as previously thought. As a result, in 2001 the US Institute of Medicine recommended a new unit, the retinol activity equivalent (RAE). Each μg RAE corresponds to 1 μg retinol, 2 μg of β-carotene in oil, 12 μg of "dietary" beta-carotene, or 24 μg of the three other dietary provitamin-A carotenoids.

Substance and its chemical environment	*Micrograms of retinol equivalent per microgram of the substance*
retinol	1
beta-carotene, dissolved in oil	1/2
beta-carotene, common dietary	1/12
alpha-carotene, common dietary	1/24
gamma-carotene, common dietary	1/24
beta-cryptoxanthin, common dietary	1/24

Because the conversion of retinol from provitamin carotenoids by the human body is actively regulated by the amount of retinol available to the body, the conversions apply strictly only for vitamin A-deficient humans. The absorption of provitamins depends greatly on the amount of lipids ingested with the provitamin; lipids increase the uptake of the provitamin.

The conclusion that can be drawn from the newer research is that fruits and vegetables are not as useful for obtaining vitamin A as was thought; in other words, the IUs that these foods were reported to contain were worth much less than the same number of IUs of fat-dissolved oils and (to some extent) supplements. This is important for vegetarians, as Night blindness is prevalent in countries where little meat or vitamin A-fortified foods are available. A sample vegan diet for one day that provides sufficient vitamin A has been published by the Food and Nutrition Board (page 120). On the other hand, reference values for retinol or its equivalents, provided by the National Academy of Sciences, have decreased. The RDA (for men) of 1968 was 5000 IU (1500 μg retinol). In 1974, the RDA was set to 1000 RE (1000 μg retinol), whereas now the Dietary Reference Intake is 900 RAE (900 μg or 3000 IU retinol). This is equivalent to 1800 μg of β-carotene supplement (3000 IU) or 10800 μg of β-carotene in food (18000 IU).

Sources

Vitamin A is found naturally in many foods:

- liver (beef, pork, chicken, turkey, fish) (6500 μg 722%), including cod liver oil
- dandelion greens (5588 IU 112%)
- carrot (835 μg 93%)
- broccoli leaf (800 μg 89%) – According to USDA database broccoli florets have much less.
- sweet potato (709 μg 79%)
- butter (684 μg 76%)
- kale (681 μg 76%)
- spinach (469 μg 52%)
- pumpkin (400 μg 41%)
- collard greens (333 μg 37%)
- Cheddar cheese (265 μg 29%)
- cantaloupe melon (169 μg 19%)
- egg (140 μg 16%)
- apricot (96 μg 11%)
- papaya (55 μg 6%)
- mango (38 μg 4%)
- pea (38 μg 4%)
- broccoli (31 μg 3%)
- milk (28 μg 3%).

Note: data taken from USDA database bracketed values are retinol activity equivalences (RAEs) and percentage of the adult male RDA, per 100 grams of the foodstuff (average).

Conversion of carotene to retinol varies from person to person and bioavailability of carotene in food varies.

Metabolic Functions

Vitamin A plays a role in a variety of functions throughout the body, such as:

- Vision
- Gene transcription
- Immune function
- Embryonic development and reproduction
- Bone metabolism
- Haematopoiesis
- Skin and cellular health
- Antioxidant activity.

Vision

The role of vitamin A in the visual cycle is specifically related to the retinal form. Within the eye, 11-*cis*-retinal is bound to rhodopsin (rods) and iodopsin (cones) at conserved lysine residues. As light enters the eye, the 11-*cis*-retinal is isomerized to the all-"trans" form. The all-"trans" retinal dissociates from the opsin in a series of steps called photo-bleaching. This isomerization induces a nervous signal along the optic nerve to the visual centre of the brain. After separating from opsin, the all-"trans"-retinal is recycled and converted back to the 11-"cis"-retinal form by a series of enzymatic reactions. In addition, some of the all-"trans" retinal may be converted to all-"trans" retinol form and then transported with an interphotoreceptor retinol-binding protein (IRBP) to the pigment epithelial cells. Further esterification into all-"trans" retinyl esters allow for storage of all-trans-retinol within the pigment epithelial cells to be reused when needed. The final stage is conversion of 11-*cis*-retinal will rebind to opsin to reform rhodopsin in the retina. Rhodopsin is needed to see in low light (contrast) as well as for night vision. It is for this reason that a deficiency in vitamin A will inhibit the reformation of rhodopsin and lead to one of the first symptoms, night blindness.

Gene Transcription

Vitamin A, in the retinoic acid form, plays an important role in gene transcription. Once retinol has been taken up by a cell, it can be oxidized to retinal (retinaldehyde) by retinol dehydrogenases and then retinaldehyde can be oxidized to retinoic acid by retinaldehyde dehydrogenases. The conversion of retinaldehyde to retinoic acid is an irreversible step, meaning that the production of retinoic acid is tightly regulated, due to its activity as a ligand for nuclear receptors. The physiological form of retinoic acid (all-trans-retinoic acid) regulates gene transcription by binding to nuclear receptors known as retinoic acid receptors (RARs) which are bound to DNA as heterodimers with retinoid "X" receptors (RXRs). RAR and RXR must dimerize before they can bind to the DNA. RAR will form a heterodimer with RXR (RAR-RXR), but it does not readily form a homodimer (RAR-RAR). RXR, on the other hand, may form a homodimer (RXR-RXR) and will form heterodimers with many other nuclear receptors as well, including the thyroid hormone receptor (RXR-TR), the Vitamin D_3 receptor (RXR-VDR), the peroxisome proliferator-activated receptor (RXR-PPAR) and the liver "X" receptor (RXR-LXR). The RAR-RXR heterodimer recognizes retinoic acid response elements (RAREs) on the DNA whereas the RXR-RXR homodimer recognizes retinoid "X" response elements (RXREs) on the DNA; although several RAREs near target genes have been shown to control physiological processes, this has not been demonstrated for RXREs. The heterodimers of RXR with nuclear receptors other than RAR (i.e. TR, VDR, PPAR, LXR) bind to various distinct response elements on the DNA to control processes not regulated by vitamin A. Upon binding of retinoic acid to the RAR component of the RAR-RXR heterodimer, the receptors undergo a conformational change that causes co-repressors to dissociate from the receptors. Coactivators can then bind to the receptor complex, which may help to loosen the chromatin structure from the histones or may interact with the transcriptional machinery. This response can upregulate (or downregulate) the expression of target genes, including Hox genes as well as the genes that encode for the receptors themselves (i.e. RAR-beta in mammals).

Dermatology

Vitamin A, and more specifically, retinoic acid, appears to maintain normal skin health by switching on genes and differentiating keratinocytes (immature skin cells) into mature epidermal cells. Exact mechanisms behind pharmacological retinoid therapy agents in the treatment of dermatological diseases are being researched. For the treatment of acne, the most prescibred retinoid drug is 13-cis retinoic acid (isotretinoin). It reduces the size and secretion of the sebaceous glands. Although it is known that 40 mg of isotretinoin will break down to an equivalent of 10 mg of ATRA — the mechanism of action of the drug (original brand name Accutane) remains unknown and is a matter of some controversy. Isotretinoin reduces bacterial numbers in both the ducts and skin surface. This is thought to be a result of the reduction in sebum, a nutrient source for the bacteria. Isotretinoin reduces inflammation via inhibition of chemotactic responses of monocytes and neutrophils. Isotretinoin also has been shown to initiate remodeling of the sebaceous glands; triggering changes in gene expression that selectively induce apoptosis. Isotretinoin is a teratogen with a number of potential side-effects. Consequently, its use requires medical supervision.

Retinal/Retinol Versus Retinoic Acid

Vitamin A deprived rats can be kept in good general health with supplementation of retinoic acid. This reverses the growth-stunting effects of vitamin A deficiency, as well as early stages of xerophthalmia. However, such rats show infertility (in both male and females) and continued degeneration of the retina, showing that these functions require retinal or retinol, which are intraconvertable but which cannot be recovered from the oxidized retinoic acid. The requirement of retinol to rescue reproduction in vitamin A deficient rats is now known to be due to a requirement for local synthesis of retinoic acid from retinol in testis and embryos.

Deficiency

Vitamin A deficiency is estimated to affect approximately one third of children under the age of five around the world. It is estimated to claim the lives of 670,000 children under five annually. Approximately 250,000–500,000 children in developing countries become blind each year owing to vitamin A deficiency, with the highest prevalence in Southeast Asia and Africa.

Vitamin A deficiency can occur as either a primary or a secondary deficiency. A primary vitamin A deficiency occurs among children and adults who do not consume an adequate intake of provitamin A carotenoids from fruits and vegetables or preformed vitamin A from animal and dairy products. Early weaning from breastmilk can also increase the risk of vitamin A deficiency.

Secondary vitamin A deficiency is associated with chronic malabsorption of lipids, impaired bile production and release, and chronic exposure to oxidants, such as cigarette smoke, and chronic alcoholism. Vitamin A is a fat soluble vitamin and depends on micellar solubilization for dispersion into the small intestine, which results in poor use of vitamin A from low-fat diets. Zinc deficiency can also impair absorption, transport, and metabolism of vitamin A because it is essential for the synthesis of the vitamin A transport proteins and as the cofactor in conversion of retinol to retinal. In malnourished populations, common low intakes of vitamin A and zinc increase the severity of vitamin A deficiency and lead physiological signs and symptoms of deficiency. A study in Burkina Faso showed major reduction of malaria morbidity with combined vitamin A and zinc supplementation in young children.

Due to the unique function of retinal as a visual chromophore, one of the earliest and specific manifestations of vitamin A deficiency is impaired vision, particularly in reduced light – night blindness. Persistent deficiency gives rise to a series of changes, the most devastating of which occur in the eyes. Some other ocular changes are referred to as xerophthalmia. First there is dryness of the conjunctiva (xerosis) as the normal lacrimal and mucus-secreting epithelium is replaced by a keratinized epithelium. This is followed by the build-up of keratin debris in small opaque plaques (Bitot's spots) and, eventually, erosion of the roughened corneal surface with softening and destruction of the cornea (keratomalacia) and total blindness. Other changes include impaired immunity (increased risk of ear infections, urinary tract infections, Meningococcal disease), hyperkeratosis (white lumps at hair follicles), keratosis pilaris and squamous metaplasia of the epithelium lining the upper respiratory passages and urinary bladder to a keratinized epithelium. With relations to dentistry, a deficiency in Vitamin A leads to enamel hypoplasia.

Adequate supply, but not excess vitamin A, is especially important for pregnant and breastfeeding women for normal fetal development. Deficiencies cannot be compensated by postnatal supplementation. Excess vitamin A, which is most common with high dose vitamin supplements, can cause birth defects and therefore should not exceed recommended daily values.

Vitamin A metabolic inhibition as a result of alcohol consumption during pregnancy is the elucidated mechanism for fetal alcohol syndrome and is characterized by teratogenicity closely matching maternal vitamin A deficiency.

Vitamin A Supplementation

Global efforts to support national governments in addressing vitamin A deficiency are led by the Global Alliance for Vitamin A (GAVA), which is an informal partnership between A2Z, the Canadian International Development Agency, Helen Keller International, the Micronutrient Initiative, UNICEF, USAID, and the World Bank. Joint GAVA activity is coordinated by the Micronutrient Initiative.

While strategies include intake of vitamin A through a combination of breast feeding and dietary intake, delivery of oral high-dose supplements remain the principal strategy for minimizing deficiency. Studies have shown vitamin A supplementation of children under five who are at risk of deficiency can reduce mortality by 23%. About 75% of the vitamin A required for supplementation activity by developing countries is supplied by the Micronutrient Initiative with support from the Canadian International Development Agency. Food fortification approaches are becoming increasingly feasible but cannot yet ensure coverage levels.

An estimated 1.25 million deaths due to vitamin A deficiency have been averted in 40 countries since 1998. In 2008 it was estimated that an annual investment of US$60 million in vitamin A and zinc supplementation combined would yield benefits of more than US$1 billion per year, with every dollar spent generating benefits of more than US$17. These combined interventions were ranked by the Copenhagen Consensus 2008 as the world's best development investment.

Toxicity

Since vitamin A is fat-soluble, disposing of any excesses taken in through diet is much harder than with water-soluble B vitamins and vitamin C, vitamin A toxicity is possible.

In general, acute toxicity occurs at doses of 25,000 IU/kg of body weight, with chronic toxicity occurring at 4,000 IU/kg of body weight daily for 6–15 months. However, liver toxicities can occur at levels as low as 15,000 IU per day to 1.4 million IU per day, with an average daily toxic dose of 120,000 IU per day, particularly with excessive consumption of alcohol. In people with renal failure, 4000 IU can cause substantial damage. In addition, excessive alcohol intake can increase toxicity. Children can reach toxic levels at 1,500 IU/kg of body weight.

Excessive vitamin A consumption can lead to nausea, irritability, anorexia (reduced appetite), vomiting, blurry vision, headaches, hair loss, muscle and abdominal pain and weakness, drowsiness, and altered mental status. In chronic cases, hair loss, dry skin, drying of the mucous membranes, fever, insomnia, fatigue, weight loss, bone fractures, anemia, and diarrhea can all be evident on top of the symptoms associated with less serious toxicity. Some of these symptoms are also common to acne treatment with Isotretinoin. Chronically high doses of vitamin A, and also pharmaceutical retinoids such as 13-cis retinoic acid, can produce the syndrome of pseudotumor cerebri. This syndrome includes headache, blurring of vision and confusion, associated with increased intracerebral pressure. Symptoms begin to resolve when intake of the offending substance is stopped.

Chronic intake of 1500 RAE of preformed vitamin A may be associated with osteoporosis and hip fractures. This may be due to the fact that an excess of vitamin A can block the expression of certain proteins dependent on vitamin K to reduce the efficacy of vitamin D, but has not yet been proven. High vitamin A intake has been associated with spontaneous bone fractures in animals. Cell culture studies have linked increased bone resorption and decreased bone formation with high intakes. This interaction may occur because vitamins A and D may compete for the same receptor and then interact with parathyroid hormone, which regulates calcium. Indeed, a study by Forsmo *et al.* shows a correlation between low bone mineral density and too high intake of vitamin A.

Toxic effects of vitamin A have been shown to significantly affect developing fetuses. Therapeutic doses used for acne treatment have been shown to disrupt cephalic neural cell activity. The fetus is particularly sensitive to vitamin A toxicity during the period of organogenesis. These toxicities only occur with preformed (retinoid) vitamin A (such as from liver). The carotenoid forms (such as beta-carotene as found in carrots), give no such symptoms, except with supplements and chronoic alcoholism, but excessive dietary intake of beta-carotene can lead to carotenodermia, which causes orange-yellow discoloration of the skin.

Smokers and chronic alcohol consumers have been observed to have increased risk of mortality due to lung cancer, esophageal cancer, gastrointestinal cancer and colon cancer. Hepatic (liver) injury been found in human and animal studies where consumption of alcohol is paired with high dose vitamin A and beta-carotene supplementation.

Researchers have succeeded in creating water-soluble forms of vitamin A, which they believed could reduce the potential for toxicity. However, a 2003 study found water-soluble vitamin A was approximately 10 times as toxic as fat-soluble vitamin. A 2006 study found children given water-soluble vitamin A and D, which are typically fat-soluble, suffer from asthma twice as much as a control group supplemented with the fat-soluble vitamins.

Vitamin A and Derivatives in Medical Use

Retinyl palmitate has been used in skin creams, where it is broken down to retinoic acid, which has potent biological activity, as described above. The retinoids, (for example, 13-cis-retinoic acid), constitute a class of chemical compounds chemically related to retinoic acid, and are used in medicine to modulate gene functions in place of this compound. Like retinoic acid, the related compounds do not have full vitamin A activity, but do have powerful effects on gene expression and epithelial cell differentiation.

Pharmaceutics utilizing mega doses of naturally occurring retinoic acid derivatives are currently in use for cancer, HIV, and dermatological purposes. At high doses, side-effects are similar to vitamin A toxicity. Severe side effects related to vitamin A toxicity, and a small optimal range of use are key obstacles in developing vitamin A-derived pharmaceutics for therapeutic use.

VITAMIN D

Vitamin D is a group of fat-soluble secosteroids. In humans vitamin D is unique both because it functions as a prohormone and because when sun exposure is adequate, the body can synthesize it (as vitamin D_3). Measures of the serum levels reflect endogenous synthesis from sun exposure as well as intake from the diet and it is believed that synthesis may contribute generally to the maintenance of adequate serum concentrations.

The evidence indicates that the synthesis of vitamin D from sun exposure works in a feedback loop that prevents toxicity but, because of uncertainty about the cancer risk from sunlight, no recommendations are issued by the Institute of Medicine for the amount of sun exposure required to meet vitamin D requirements.

Accordingly the Dietary Reference Intakes for vitamin D assume that no synthesis occurs and that all of a person's vitamin D is from his or her diet. When synthesized in the kidneys, calcitriol circulates as a hormone, regulating the concentration of calcium and phosphate in the bloodstream and promoting the healthy growth and remodeling of bone.

Vitamin D prevents rickets in children and osteomalacia in adults, and, together with calcium, helps to protect older adults from osteoporosis. Vitamin D also affects neuromuscular function, inflammation, and influences the action of many genes that regulate the proliferation, differentiation and apoptosis of cells.

Forms

Several forms (vitamers) of vitamin D exist. The two major forms are vitamin D_2 or ergocalciferol, and vitamin D_3 or cholecalciferol, vitamin D without a subscript refers to either D_2 or D_3 or both. These are known collectively as calciferol. Vitamin D_2 was chemically characterized in 1932. In 1936, the chemical structure of vitamin D_3 was established and resulted from the ultraviolet irradiation of 7-dehydrocholesterol.

Chemically, the various forms of vitamin D are secosteroids; i.e., steroids in which one of the bonds in the steroid rings is broken. The structural difference between vitamin D_2 and vitamin D_3 is in their side chains. The side chain of D_2 contains a double bond between carbons 22 and 23, and a methyl group on carbon 24.

Vitamin D_3 (cholecalciferol) is produced by ultraviolet irradiation (UV) of its precursor 7-dehydrocholesterol. This molecule occurs naturally in the skin of animals and in milk. Vitamin D_3 can be made by exposure of the skin to UV, or by exposing milk directly to UV (one commercial method).

Vitamin D_2 is a derivative of ergosterol, a membrane sterol named for the ergot fungus, which is produced by some organisms of phytoplankton, invertebrates, and fungi. The vitamin ergocalciferol (D_2) is produced in these organisms from ergosterol in response to UV irradiation. D_2 is not produced by land plants or vertebrates, because they lack the precursor ergosterol. The biological fate for producing 25(OH)D from vitamin D_2 is expected to be the same as for D_3, although some controversy exists over whether or not D_2 can fully substitute for vitamin D_3 in the human diet.

EVOLUTION

The photosynthesis of vitamin D evolved over an estimated 750 million years ago; the phytoplankton coccolithophore *Emiliania huxleyi* is an early example. Vitamin D played a critical role in the maintenance of a calcified skeleton in vertebrates as they left their calcium-rich ocean environment for land over an estimated 350 million years ago.

Vitamin D can be synthesized only via a photochemical process, so early vertebrates that ventured onto land either had to ingest foods that contained vitamin D or had to be exposed to sunlight to photosynthesize vitamin D in their skin to satisfy their body's vitamin D requirement.

Production in the Skin

Vitamin D_3 is made in the skin when 7-dehydrocholesterol reacts with ultraviolet light (UVB) at wavelengths between 270 and 300 nm, with peak synthesis occurring between 295 and 297 nm. These wavelengths are present in sunlight when the UV index is greater than three and also in the light emitted by the UV lamps in tanning beds. Tanning lamps produce ultraviolet primarily in the UVA spectrum, but typically produce 4% to 10% of the total UV emissions as UVB. At this solar elevation, which occurs daily within the tropics, daily during the spring and summer seasons in temperate regions, and almost never within the arctic circles, vitamin D_3 can be made in the skin. Latitude does not consistently predict the average serum 25OHD level of a population. The assumption that vitamin D levels in the population follow a latitude gradient is especially questionable in view of surveys which have shown that UVB penetrating to the earth's surface over 24 hours during the summer months in northern Canada equals or exceeds UVB penetration at the equator. Accordingly, there is sufficient opportunity during the spring, summer, and fall months at high latitude for humans to form and store vitamin D_3. Depending on the intensity of UVB rays and the minutes of exposure, an equilibrium can develop in the skin, and vitamin D degrades as fast as it is generated.

The skin consists of two primary layers: the inner layer called the dermis, composed largely of connective tissue, and the outer, thinner epidermis. Thick epidermis in the soles and palms consists of five *strata*; from outer to inner they are: the stratum corneum, stratum lucidum, stratum granulosum, stratum spinosum, and stratum basale. Vitamin D is produced in the two innermost strata, the stratum basale and stratum spinosum.

Cholecalciferol is produced photochemically in the skin from 7-dehydrocholesterol; 7-dehydrocholesterol is produced in relatively large quantities in the skin of most vertebrate animals, including humans. The naked mole rat appears to be naturally cholecalciferol deficient, as serum 25-OH vitamin D levels are undetectable. In some animals, the presence of fur or feathers blocks the UV rays from reaching the skin. In birds and fur-bearing mammals, vitamin D is generated from the oily secretions of the skin deposited onto the feathers or fur and is obtained orally during grooming.

Discovery

American researchers Elmer McCollum and Marguerite Davis in 1913 discovered a substance in cod liver oil which later was called "vitamin A". British doctor Edward Mellanby noticed dogs that were fed cod liver oil did not develop rickets and concluded vitamin A, or a closely associated factor, could prevent the disease. In 1921, Elmer McCollum tested modified cod liver oil in which the vitamin A had been destroyed. The modified oil cured the sick dogs, so McCollum concluded the factor in cod liver oil which cured rickets was distinct from vitamin A. He called it vitamin D because it was the fourth vitamin to be named. It was not initially realized that, unlike other vitamins, vitamin D can be synthesised by humans through exposure to UV light.

In 1923, it was established that when 7-dehydrocholesterol is irradiated with light, a form of a fat-soluble vitamin is produced (now known as D_3). Alfred Fabian Hess showed "light equals vitamin D." Adolf Windaus, at the University of Göttingen in Germany, received the Nobel Prize in Chemistry in 1928, for his work on the constitution of sterols and their connection with vitamins. In the 1930s he clarified further the chemical structure of vitamin D.

In 1923, Harry Steenbock at the University of Wisconsin demonstrated that irradiation by ultraviolet light increased the vitamin D content of foods and other organic materials. After irradiating rodent food, Steenbock discovered the rodents were cured of rickets. A vitamin D deficiency is a known cause of rickets. Using $300 of his own money, Steenbock patented his invention. His irradiation technique was used for foodstuffs, most memorably for milk. By the expiration of his patent in 1945, rickets had been all but eliminated in the US.

Industrial Production

Vitamin D_3 (cholecalciferol) is produced industrially by exposing 7-dehydrocholesterol to UVB light, followed by purification. The 7-dehydrocholesterol is a natural substance in wool grease (lanolin) from sheep or other woolly animals. Vitamin D2 (ergocalciferol) is produced in a similar way using ergosterol from yeast or mushrooms as a starting material.

Mechanism of Action

Vitamin D is carried in the bloodstream to the liver, where it is converted into the prohormone calcidiol. Circulating calcidiol may then be converted into calcitriol, the biologically active form of vitamin D, either in the kidneys or by monocyte-macrophages in the immune system. When synthesized by monocyte-macrophages, calcitriol acts locally as a cytokine, defending the body against microbial invaders. Following the final converting step in the kidney, calcitriol (the physiologically active form of vitamin D) is released into the circulation. By binding to vitamin D-binding protein (VDBP), a carrier protein in the plasma, calcitriol is transported to various target organs.

Calcitriol mediates its biological effects by binding to the vitamin D receptor (VDR), which is principally located in the nuclei of target cells. The binding of calcitriol to the VDR allows the VDR to act as a transcription factor that modulates the gene expression of transport proteins (such as TRPV6 and calbindin), which are involved in calcium absorption in the intestine. The vitamin D receptor belongs to the nuclear receptor superfamily of steroid/thyroid hormone receptors, and VDRs are expressed by cells in most organs, including the brain, heart, skin, gonads, prostate, and breast. VDR activation in the intestine, bone, kidney, and parathyroid gland cells leads to the maintenance of calcium and phosphorus levels in the blood (with the assistance of parathyroid hormone and calcitonin) and to the maintenance of bone content.

Vitamin D increases expression of the tyrosine hydroxylase gene in adrenal medullary cells. It also is involved in the biosynthesis of neurotrophic factors, synthesis of nitric oxide synthase, and increased glutathione levels.

The VDR is known to be involved in cell proliferation and differentiation. Vitamin D also affects the immune system, and VDRs are expressed in several white blood cells, including monocytes and activated T and B cells.

Apart from VDR activation, various alternative mechanisms of action are known. An important one of these is its role as a natural inhibitor of signal transduction by hedgehog (a hormone involved in morphogenesis).

Vitamin D and Health Outcomes

The Institute of Medicine report *Dietary Reference Intakes for Calcium and Vitamin D* was requested by the U.S. and Canadian governments. The report was the work of an ad hoc consensus committee of 14 scientists with "expertise in the areas of vitamin D and calcium or a related topic area, with specific expertise related to pregnancy and reproductive nutrition, pediatrics and infant nutrition, minority health and health disparities, cellular metabolism, toxicology and risk assessment, dermatology, immunology, endocrinology, skeletal health, oncology, cardiovascular health, epidemiology; nutrition monitoring, and biostatistics". The report examined literature identified by the committee and incorporated systematic evidence-based reviews from the Agency for Healthcare Research and Quality to assess indicators of chronic and non-chronic disease outcomes.

The Institute of Medicine's report provides the most objective scientifically based information about what is adequate or excess intake of vitamin D and the role of vitamin D in human health.

"Outcomes related to cancer/neoplasms, cardiovascular disease and hypertension, diabetes and metabolic syndrome, falls and physical performance, immune functioning and autoimmune disorders, infections, neuropsychological functioning, and preeclampsia could not be linked reliably with calcium or vitamin D intake and were often conflicting. "

According to the report: "Despite the many claims of benefit surrounding vitamin D in particular, the evidence did not support a basis for a causal relationship between vitamin D and many of the numerous health outcomes purported to be affected by vitamin D intake. Although the current interest in vitamin D as a nutrient with broad and expanded benefits is understandable, it is not supported by the available evidence. The established function of vitamin D remains that of ensuring bone health, for which causal evidence across the life stages exists and has grown since the 1997 DRIs were established (IOM, 1997). The conclusion that there is not sufficient evidence to establish a relationship between vitamin D and health outcomes other than bone health does not mean that future research will not reveal a compelling relationship between vitamin D and another health outcome. The question is open as to whether other relationships may be revealed in the future."

Serum 25-Hydroxyvitamin D

The Institute of Medicine committee concluded that a serum 25-hydroxyvitamin D level of 20 ng/mL is desirable for bone and overall health. The Dietary Reference Intakes for vitamin D are chosen with a margin of safety and 'overshoot' the targeted serum value to ensure that the specified levels of intake achieve the desired serum 25-hydroxyvitamin D levels in almost all persons. It is assumed there are no contributions to serum 25-hydroxyvitamin D level from sun exposure and the recommendations are fully applicable to people with dark skin or negligible exposure to sunlight. The Institute of Medicine found that serum 25-hydroxyvitamin D concentrations above 30 ng/mL are "not consistently associated with increased benefit". Serum 25-hydroxyvitamin D levels above 50 ng/mL may be cause for concern.

Dietary Reference Intakes

The Dietary Reference Intake for vitamin D issued by the Institute of Medicine in 2010 superseded a previous recommendation which had Adequate Intake status. The recommendations were formed assuming the individual has no skin synthesis of vitamin D because of inadequate sun exposure. The reference intake for vitamin D refers to total intake from food, beverages and supplements, is intended for the North American population, and assumes that calcium requirements are being met.

The new reference intakes for vitamin D are:

- 1–70 years of age: 600 IU/day (15 μg equivalent)
- 71+ years of age: 800 IU/day
- Pregnant/lactating: 600 IU/day

The reference AI for infants remains at:

- 0–12 months: 400 IU/day

Potential for Adverse Effects from Excess Vitamin D

Vitamin D toxicity has never been reported from ultraviolet B light alone. The Institute of Medicine used a risk assessment framework in reviewing all available data related to vitamin D intake and various health outcomes. The aim was not to prove that certain levels of intake certainly cause harm, but rather to decide whether the emerging evidence warranted caution relative to vitamin D intakes and associated serum 25-hydroxyvitamin D concentrations less than those associated with acute toxicity but still associated with adverse effects that may occur as a result of chronic intake.

Tolerable Upper Intake Levels

The Tolerable Upper Intake Level is defined as "the highest average daily intake of a nutrient that is likely to pose no risk of adverse health effects for nearly all persons in the general population". Although tolerable upper intake levels are believed to be safe, information on the long term effects is incomplete and these levels of intake are not recommended :

- 0–6 months of age: 1,000 IU
- 6–12 months of age: 1,500 IU
- 1–3 years of age: 2,500 IU
- 4–8 years of age: 3,000 IU
- 9-71+ years of age: 4,000 IU
- Pregnant/lactating: 4,000 IU.

Cardiovascular Disease

A report from the National Health and Nutrition Examination Survey (NHANES) involving nearly 5,000 participants found that low levels of vitamin

D were associated with an increased risk of peripheral artery disease (PAD). The incidence of PAD was 80% higher in participants with the lowest vitamin D levels (<17.8 ng/mL). Cholesterol levels were found to be reduced in gardeners in the UK during the summer months. Low levels of vitamin D are associated with an increase in high blood pressure and cardiovascular risk. Numerous observational studies show this link, but of two systemic reviews one found only weak evidence of benefit from supplements and the other found no evidence of a beneficial effect whatsoever.

There is a certain amount of evidence to suggest that dietary vitamin D may be carried by lipoprotein particles into cells of the artery wall and atherosclerotic plaque, where it may be converted to active form by monocyte-macrophages. These findings raise questions regarding the effects of vitamin D intake on atherosclerotic calcification and cardiovascular risk. Calcifediol (25-hydroxy-vitamin D) is implicated in the etiology of atherosclerosis, especially in non-Caucasians. Freedman et al. (2010) found that serum vitamin D correlates with calcified atheroscleratic plaque (CP) in African Americans, but not in Euro-Americans, "Higher levels of 25-hydroxyvitamin D seem to be positively associated with aorta and carotid CP in African Americans but not with coronary CP. These results contradict what is observed in individuals of European descent." One study found an elevated risk of ischaemic heart disease in Southern India in individuals whose vitamin D levels were above 89 ng/mL. A review of vitamin D status in India concluded that studies uniformly point to low 25(OH)D levels in Indians despite abundant sunshine, and suggested a public health need to fortify Indian foods with vitamin D might exist. The levels found in India are consistent with many other studies of tropical populations which have found that even an extreme amount of sun exposure, such as incurred by rural Indians, does not raise 25(OH)D levels to the levels typically found in Europeans.

Mortality

Using information from the National Health and Nutrition Examination Survey a large scale study concluded that having low levels of vitamin D (<17.8 ng/ml) was independently associated with an increase in all-cause mortality in the general population. However it has been pointed out that increased mortality was also found in those with higher concentrations, (above 50 ng/ml). A sophisticated August 2010 study of plasma vitamin D and mortality in older men concluded that both high (>39 ng/ml) and low (<18 ng/ml) concentrations of plasma 25(OH)D are associated with elevated risks of overall and cancer mortality compared with intermediate concentrations. These boundaries were less than suggested by the Melamed et al. study of National Health and Nutrition Examination Survey data but the immunoassay used by National Health and Nutrition Examination Survey tended to overestimate vitamin D values. Overall, excess or deficiency in the calciferol system appear to cause abnormal functioning and premature aging. The data suggest a U-shaped shaped risk curve between serum 25OHD level and all-cause mortality; increases in risk with high levels appear at a lower threshold for the black population.

Complex regulatory mechanisms control metabolism and recent epidemiological evidence suggests that there is a narrow range of vitamin D blood levels in which metabolic functions are optimized. Levels above or below this natural homeostasis of vitamin D are associated with increased mortality.

Vitamin E

Vitamin E or tocopherol is a fat-soluble vitamin. It is stable to heat and acids. It is rapidly oxidized in rancid fats. Many claims are being made that supplements of this vitamin can prevent or cure a wide variety of diseases, from reproductive function to skin problems such as psoriasis and acne, but there is no proof.

Functions Vitamin E is the most potent natural antioxidant found in food. Polyunsaturated fatty acids form a structural part of all cell membranes. They are prone to oxidative breakdown by free radicals in the cell. The main function of vitamin E is to act as an antioxidant. Vitamin E itself gets oxidized and protects cell membranes from oxidative damage. It performs the following functions:

1. Prevents oxidation of vitamin A in the intestine
2. Protects normal cell membranes by preventing their breakdown 3. Prevents hemolysis of red blood cells
4. Prevents oxidation of PUFAs.

Sources Vitamin E is widely distributed in foods, particularly vegetable oils (corn, soya, sunflower, safflower), wheat germ, whole grains, legumes, nuts, and dark green leafy vegetables.

Deficiency In severe deficiency, although uncommon, increased haemolysis of red blood cells is seen in premature infants.

Vitamin K

Vitamin K is essential in the diet because it is needed for synthesis of prothrombin and other blood clotting factors. It exists as K_1 (found in plants), K_2 (synthesized by bacteria in the intestinal tract), and K_3 (synthetic form). Being fat soluble, it requires fat and bile salts for efficient absorption.

Functions Vitamin K is required for the formation of prothrombin and several other proteins involved in clotting of blood. The ability of blood to clot is dependent upon a high blood level of prothrombin. Deficiency A deficiency of vitamin K is uncommon in adults. New born infants have a sterile intestinal tract, hence they are given a single dose of vitamin K to prevent haemorrhagic disease. A deficiency interferes with formation of prothrombinogen and, thus, reduces clotting tendency of blood. It may occur during diseases of malabsorption, oral use of sulfa drugs and antibiotics, or certain drugs which are vitamin K antagonists and can cause haemorrhages.

Sources Bacterial synthesis in the intestinal tract supplies at least half of the daily needs. Green leafy vegetables, cabbage, cauliflower, and pork liver are excellent sources. Cheese, egg yolk, and tomato also supply vitamin K.

VITAMIN DEFICIENCY

VITAMIN A

Vitamin A deficiency is a lack of vitamin A in humans. It is common in developing countries but rarely seen in developed countries. Night blindness is one of the first signs of vitamin A deficiency. Xerophthalmia and complete blindness can also occur since Vitamin A has a major role in phototransduction. Approximately 250,000 to 500,000 malnourished children in the developing world go blind each year from a deficiency of vitamin A, approximately half of which die within a year of becoming blind.

The United Nations Special Session on Children in 2002 set the elimination of vitamin A deficiency by 2010. The prevalence of night blindness due to vitamin A deficiency is also high among pregnant women in many developing countries. Vitamin A deficiency also contributes to maternal mortality and other poor outcomes in pregnancy and lactation.

Vitamin A deficiency also diminishes the ability to fight infections. In countries where children are not immunized, infectious disease like measles have higher fatality rates. As elucidated by Dr. Alfred Sommer, even mild, subclinical deficiency can also be a problem, as it may increase children's risk of developing respiratory and diarrheal infections, decrease growth rate, slow bone development, and decrease likelihood of survival from serious illness.

SIGNS AND SYMPTOMS

The most common cause of blindness in developing countries is vitamin A deficiency (VAD). The World Health Organization (WHO) estimates 13.8 million children to have some degree of visual loss related to VAD. Night blindness and its worsened condition, xerophthalmia, are markers of VAD, as VAD can also lead to impaired immune function, cancer, and birth defects.

Night blindness is the difficulty for the eyes to adjust to dim light. Affected individuals are unable to distinguish images in low levels of illumination. People with night blindness have poor vision in the darkness, but see normally when adequate light is present. VAD affects vision by inhibiting the production of rhodopsin, the eye pigment responsible for sensing low light situations. Rhodopsin is found in the retina and is composed of retinal (an active form of vitamin A) and opsin (a protein). Because the body cannot create retinal in sufficient amounts, a diet low in vitamin A will lead to a decreased amount of rhodopsin in the eye, as there is inadequate retinal to bind with opsin. Night blindness results.

Night blindness caused by VAD has been associated with the loss of goblet cells in the conjunctiva, a membrane covering the outer surface of the eye. Goblet cells are responsible for secretion of mucus, and their absence results in xerophthalmia, a condition where the eyes fail to produce tears. Dead epithelial and microbial cells accumulate on the conjunctiva and form debris that can lead to infection and possibly blindness.

Decreasing night blindness requires the improvement of vitamin A status in at risk populations. Supplements and fortification of food have been shown to be effective interventions. Supplement treatment for night blindness includes high doses of vitamin A (200,000 IU) in the form of retinyl palmitate to be taken by mouth, which is administered two to four times a year. Intramuscular injections are poorly absorbed and are ineffective in delivering sufficient bio-available vitamin A. Fortification of food with vitamin A is costly, but can be done in wheat, sugar, and milk. Households may circumvent expensive fortified food by altering dietary habits. Consumption of yellow-orange fruits and vegetables rich in carotenoids, specifically beta carotene, provides pro-vitamin A precursors that will prevent VAD related night blindness.

Causes

The major cause is diets which include few animal sources of pre-formed vitamin A. Breast milk of a lactating mother with vitamin A deficiency contains little vitamin A, which provides a breast-fed child with too little vitamin A.

In addition to dietary problems, there are other causes of vitamin A deficiency. Iron deficiency can affect vitamin A uptake. Excess alcohol consumption can deplete vitamin A, and a stressed liver may be more susceptible to vitamin A toxicity. People who consume large amounts of alcohol should seek medical advice before taking vitamin A supplements. In general, people should also seek medical advice before taking vitamin A supplements if they have any condition associated with fat malabsorption such as pancreatitis, cystic fibrosis, tropical sprue & biliary obstruction.

Treatment

Treatment of vitamin A deficiency can be undertaken with both oral and injectable forms, generally as vitamin A palmitate.

As an oral form, the supplementation of vitamin A is effective for lowering the risk of morbidity, especially from severe diarrhea, and reducing mortality from measles and all-cause mortality. Some countries where vitamin A deficiency is a public health problem address its elimination by including vitamin A supplements available in capsule form with National Immunization Days (NIDs) for polio eradication or measles.

Vitamin A capsules cost about US$0.02. The capsules are easy to handle; they don't need to be stored in a refrigerator or vaccine carrier. When the correct dosage is given, vitamin A is safe and has no negative effect on seroconversion rates for Oral Polio Vaccine or measles vaccine.

However, because the benefit of vitamin A supplements is transient, children need them regularly every four to six months. Since NIDs provide only one dose per year, NIDs-linked vitamin A distribution must be complemented by other dose programs to maintain vitamin A in children Maternal high supplementation benefits both mother and breast-fed infant: high dose vitamin A supplementation of the lactating mother in the first month postpartum can provide the breast-fed infant with an appropriate amount of vitamin A through breast milk. However, high-dose supplementation of pregnant women should be avoided because it can cause miscarriage and birth defects.

Some non-profit, non-governmental organizations have taken on the task of Vitamin A deficiency prevention as well. One such organization is Vitamin Angels. Vitamin Angels has committed itself to eradicating childhood blindness due to Vitamin A deficiency on the planet by the year 2020. Operation 20/20 was launched in 2007 and will cover 18 countries. The program gives children two high dose vitamin A and anti-parasitic supplements (twice a year for four years), which provides children with enough of the nutrient during their most vulnerable years in order to prevent them from going blind and suffering from other life-threatening diseases caused by Vitamin A Deficiency.

Food fortification is also useful for improving vitamin A deficiency. A variety of oily and dry forms of the retinol esters, retinyl acetates and retinyl palmitate are available for food fortification of vitamin A. Margarine and oil are the ideal food vehicles for vitamin A fortification. They protect vitamin A from oxidation during storage and prompt absorption of vitamin A. â-carotene and retinyl acetate or retinyl palmitate are used as a form of vitamin A for vitamin A fortification of fat-based foods. Fortification of sugar with retinyl palmitate as a form of vitamin A has been used extensively throughout

Central America. Cereal flours, milk powder, and liquid milk are also used as food vehicles for vitamin A fortification.

Dietary diversification can also control vitamin A deficiency. Non-animal sources of vitamin A which contain pre-formed vitamin A account for greater than 80% of intake for most individuals in the developing world. The increase in consumption of vitamin A-rich foods of animal origin in addition to fruits and vegetables has beneficial effects on vitamin A deficiency.

VITAMIN E

Vitamin E deficiency causes neurological problems due to poor nerve conduction. These include neuromuscular problems such as spinocerebellar ataxia and myopathies. Deficiency can also cause anemia, due to oxidative damage to red blood cells.

Vitamin E deficiency is very rare in humans and is almost never caused by a poor diet. Instead, there are three specific situations when a vitamin E deficiency is likely to occur. It is seen in persons who cannot absorb dietary fat, has been found in premature, very low birth weight infants (birth weights less than 1500 grams, or 3.5 pounds), and is seen in individuals with rare disorders of fat metabolism.

Individuals who cannot absorb fat may require a vitamin E supplement because some dietary fat is needed for the absorption of vitamin E from the gastrointestinal tract. Anyone diagnosed with cystic fibrosis, individuals who have had part or all of their stomach removed, and individuals with malabsorptive problems such as Crohn's disease, liver disease or pancreatic insufficiency may not absorb fat and should discuss the need for supplemental vitamin E with their physician. People who cannot absorb fat often pass greasy stools or have chronic diarrhea and bloating. Very low birth weight infants may be deficient in vitamin E. A neonatologist, a pediatrician specializing in the care of newborns, typically evaluates the nutritional needs of premature infants.

Abetalipoproteinemia is a rare inherited disorder of fat metabolism that results in poor absorption of dietary fat and vitamin E. The vitamin E deficiency associated with this disease causes problems such as poor transmission of nerve impulses, muscle weakness, and degeneration of the retina that can cause blindness. Individuals with abetalipoproteinemia may be prescribed special vitamin E supplements by a physician to treat this disorder. In addition, there is a rare genetic condition termed isolated vitamin E deficiency or ataxia with isolated with vitamin E deficiency, caused by mutations in the gene for the tocopherol transfer protein. These individuals have an extremely poor capacity to absorb vitamin E and develop neurological complications that are reversed by high doses of vitamin E.

VITAMIN K DEFICIENCY

Causes

Vitamin K-deficiency may occur by disturbed intestinal uptake (such as would occur in a bile duct obstruction), by therapeutic or accidental intake of vitamin K-antagonists or, very rarely, by nutritional vitamin K deficiency. As a result, Gla-residues are inadequately formed and the Gla-proteins are insufficiently active.

Symptoms

Symptoms include ecchymosis, petechiae, hematomas, oozing of blood at surgical or puncture sites, stomach pains; risk of massive uncontrolled bleeding; cartilage calcification; and severe malformation of developing bone or deposition of insoluble calcium salts in the walls of arteries.

Vitamin Supplementation

According to a study published in the October 14, 2008 edition of PLoS Medicine, Vitamin K (5mg of K_1/day) does not protect against age-related decreasing bone density, but may protect against fractures and cancers, in postmenopausal women taking calcium and vitamin D supplements.

Menaquinone (vitamin K_2), but not phylloquinone (vitamin K_1), intake is associated with reduced risk of CHD mortality, all-cause mortality and severe aortic calcification.

In a cohort study in Germany (11319 men, mean follow-up time 8.6y), Menaquinone intake was associated with decreased incidence of advanced prostate cancer.

Prevalence

The prevalence of vitamin K deficiency varies by geographic region.

For infants in the United States, vitamin K deficiency without bleeding may occur in as many as 50% of infants younger than 5 days old. Therefore, the Committee on Nutrition of the American Academy of Pediatrics recommends that 0.5 to 1.0 mg Vitamin K1 be administered to all newborns shortly after birth.

Postmenopausal and elderly women in Thailand have high risk of Vitamin K_2 deficiency, compared with the normal value of young, reproductive females. Current dosage recommendations for Vitamin K may be too low.

The deposition of calcium in soft tissues, including arterial walls, is quite common, especially in those suffering from atherosclerosis, suggesting that Vitamin K deficiency is more common than previously thought.

Kwashiorkor

A seated, listless child, who was among many kwashiorkor cases found in Nigerian relief camps during the Nigerian–Biafran War.

Kwashiorkor is an acute form of childhood protein-energy malnutrition characterized by edema, irritability, anorexia, ulcerating dermatoses, and an enlarged liver with fatty infiltrates. The presence of edema caused by poor nutrition defines kwashiorkor. Kwashiorkor was thought to be caused by insufficient protein consumption but with sufficient calorie intake, distinguishing it from marasmus. More recently, micronutrient and antioxidant deficiencies have come to be recognized as contributory. Cases in the developed world are rare.

Jamaican pediatrician Dr. Cicely D. Williams introduced the name into the medical community in her 1935 Lancet article. The name is derived from the Ga language of coastal Ghana, translated "the sickness the baby gets when the new baby comes", and reflecting the development of the condition in an older child who has been weaned from the breast when a younger sibling comes. Breast milk contains proteins and amino acids vital to a child's growth. In at-risk populations, kwashiorkor may develop after a mother weans her child from breast milk and replaces the diet with foods high in starches and carbohydrates and deficient in protein.

Signs and Symptoms

The defining sign of kwashiorkor in a malnourished child is pedal edema (swelling of the feet). Other signs include a distended abdomen, an enlarged liver with fatty infiltrates, thinning hair, loss of teeth, skin depigmentation and dermatitis. Children with kwashiorkor often develop irritability and anorexia.

Victims of kwashiorkor fail to produce antibodies following vaccination against diseases, including diphtheria and typhoid. Generally, the disease can be treated by adding food energy and protein to the diet; however, it can have a long-term impact on a child's physical and mental development, and in severe cases may lead to death.

Possible Causes

There are various explanations for the development of kwashiorkor and the topic remains controversial. It is now accepted that protein deficiency, in combination with energy and micronutrient deficiency, is necessary but not sufficient to cause kwashiorkor. The condition is likely due to deficiency of one of several types of nutrients (e.g., iron, folic acid, iodine, selenium, vitamin C), particularly those involved with anti-oxidant protection. Important anti-oxidants in the body that are reduced in children with kwashiorkor include glutathione, albumin, vitamin E and polyunsaturated fatty acids. Therefore, if a child with reduced type one nutrients or anti-oxidants is exposed to stress (e.g. an infection or toxin) he/she is more liable to develop kwashiorkor.

Ignorance of nutrition can be a cause. Dr. Latham, director of the Program in International Nutrition at Cornell University cited a case where parents who fed their child cassava failed to recognize malnutrition because of the edema caused by the syndrome and insisted the child was well-nourished despite the lack of dietary protein.

One important factor in the development of kwashiorkor is aflatoxin poisoning. Aflatoxins are produced by molds and ingested with moldy foods. They are toxified by the cytochrome P450 system in the liver, the resulting epoxides damage liver DNA. Since many serum proteins, in particular albumin, are produced in the liver, the symptoms of kwashiorkor are easily explained. It is noteworthy that kwashiorkor occurs mostly in warm, humid climates that encourage mold growth. In dry climates, marasmus is the more frequent disease associated with malnutrition. This has important consequences for treatment of the patients. Protein should be supplied only for anabolic purposes. The catabolic needs should be satisfied with carbohydrate and fat. Protein catabolism involves the urea cycle, which is located in the liver and can easily overwhelm the capacity of an already damaged organ. The resulting liver failure can be fatal.

Marasmus

Marasmus is a form of severe protein-energy malnutrition characterized by energy deficiency.

A child with marasmus looks emaciated. Body weight may be reduced to less than 80% of the normal weight for that height. Marasmus occurrence increases prior to age 1, whereas kwashiorkor occurrence increases after 18 months. The prognosis is better than it is in kwashiorkor.

Signs and Symptoms

The malnutrition associated with marasmus leads to extensive tissue and muscle wasting, as well as variable edema. Other common characteristics include dry skin, loose skin folds hanging over the glutei, axillae, etc. There is also drastic loss of adipose tissue from normal areas of fat deposits like buttocks and thighs. The afflicted are often fretful, irritable, and voraciously hungry.

Causes

Marasmus is caused by a severe deficiency of nearly all nutrients, especially protein and calories.

Treatment

It is necessary to treat not only the symptoms but also the complications of the disorder, including infections, dehydration, and circulation disorders, which are frequently lethal and lead to high mortality if ignored. Ultimately, marasmus can progress to the point of no return when the body's machinery for protein synthesis, itself made of protein, has been degraded to the point that it cannot handle any protein. At this point, attempts to correct the disorder by giving food or protein are futile.

Catabolysis

Catabolysis is a biological process in which the body breaks down fat and muscle tissue in order to stay alive. Catabolysis occurs only when there is no longer any source of protein, carbohydrate, or vitamin nourishment feeding all body systems; it is the most severe type of malnutrition.

Disease Settings

Catabolysis is seen in starved and malnourished people. Over 20,000 people die from starvation each day. Also, officials estimate that over 800,000,000 people are "chronically undernourished" and that a child dies from starvation every five seconds.

Mechanism

Due to the normal metabolic rate of humans (which requires going approximately 12 hours without food), catabolysis becomes life-threatening only after 1–2 months from the cessation of nutrition going into the body. After this time, the damage to muscles and organs can be permanent and can also eventually cause death, if left untreated. Catabolysis is a last resort effort of the body to keep itself — particularly the nervous system — alive.

Protein stores, especially in muscle tissue, provide the amino acids needed for the process. Amino acids are released into the blood and converted in the liver to alpha keto acids. Alpha keto acids can than be converted to glucose to maintain proper blood sugar levels.

The situation can become dire when one begins to lose muscle mass; this is a sign that the fat has been expended and the body is now metabolizing the muscle tissue. This results in muscle atrophy, a loss of strength and, ultimately, a depletion of muscular tissue completely. Muscle weakness is not necessarily a symptom of catabolysis: the muscles will normally feel fatigued when they are not receiving enough energy or oxygen. Ultimately, catabolysis can progress to the point of no return when the body's machinery for protein synthesis, itself made of protein, has been degraded to the point that it cannot handle any protein. At this point, attempts to correct the disorder by giving food or protein are futile. The body has a natural store of fat (also called *adipose*) that stores reserve energy. One can still stay alive while the body breaks down the fatty tissue (hence people wasting away from starvation). The person may, during catabolysis, have large amounts of lipids, proteins, and amino acids in the bloodstream, due to the muscle fibers' and adipose tissue's being broken down and sent to the nervous system and brain. One may also exhibit a fever, since the body is working hard to transfer the nutrients in the muscles and fat to the blood.

Treatment

While catabolysis can be deadly over time, if the person is given medical treatment early enough, the effects of catabolysis can be reversed. However, the person may require intravenous nutrition, a blood transfusion, and/or oxygen replenishment. After that, it may be a few weeks to a few months before the person's muscle mass and fat deposits can build themselves up again; there is a possibility that they may never build back up, depending on the severity of the condition.

Thiamine

Thiamine or thiamin or Vitamin B1, and named as the "thio-vitamine" ("sulfur-containing vitamin") is a water-soluble vitamin of the B complex. First named *aneurin* for the detrimental neurological effects of its lack in the diet, it was eventually assigned the generic descriptor name vitamin B_1. Its phosphate derivatives are involved in many cellular processes. The best characterized form is thiamine pyrophosphate (ThDP), a coenzyme in the catabolism of sugars and amino acids. In yeast, ThDP is also required in the first step of alcoholic fermentation.

All living organisms use thiamine in their biochemistry, but it is synthesized in bacteria, fungi and plants. Animals must obtain it from their diet, and thus for them it is a vitamin.

Insufficient intake in birds produces a characteristic polyneuritis, and in mammals results in a disease called beriberi affecting the peripheral nervous system (polyneuritis) and/or the cardiovascular system, with fatal outcome if not cured by thiamine administration.

In less severe deficiency, nonspecific signs include malaise, weight loss, irritability and confusion.

Today, there is still a much work devoted to elucidating the exact mechanisms by which thiamine deficiency leads to the specific symptoms observed. New thiamine phosphate derivatives have recently been discovered, emphasizing the complexity of thiamine metabolism and the need for more research in the field.

DIETARY MINERAL

Dietary minerals (also known as mineral nutrients) are the chemical elements required by living organisms, other than the four elements carbon, hydrogen, nitrogen, and oxygen present in common organic molecules.

Examples of mineral elements include calcium, magnesium, potassium, sodium, zinc, and iodine. Most minerals that enter into the dietary physiology of organisms consist of simple chemical elements.

Larger aggregates of minerals need to be broken down for absorption. Bacteria play an essential role in the weathering of primary minerals that results in the release of nutrients for their own nutrition and for the nutrition of others in the ecological food chain.

Scientists are only recently starting to appreciate the magnitude and role that microorganisms have in the global cycling and formation of biological minerals.

Plants absorb dissolved minerals in soils, which are subsequently picked up by the herbivores that eat them and so on, the minerals move up the food chain. Larger organisms may also consume soil (geophagia) and visit mineral licks to obtain limiting mineral nutrients they are unable to acquire through other components of their diet.

Essential Chemical Elements

Some sources state that sixteen chemical elements are *required* to support human biochemical processes by serving structural and functional roles as well as electrolytes: Most of the dietary elements are of relatively low atomic weight:

Dietary Nutrition

Dietitians may recommend that dietary elements are best supplied by ingesting specific foods rich with the chemical element(s) of interest.

The elements may be naturally present in the food (e.g., calcium in dairy milk) or added to the food (e.g., orange juice fortified with calcium; iodized salt, salt fortified with iodine).

Dietary supplements can be formulated to contain several different chemical elements (as compounds), a combination of vitamins and/or other chemical compounds, or a single element (as a compound or mixture of compounds), such as calcium (as carbonate, citrate, etc.) or magnesium (as oxide, etc.), chromium (usually as picolinate) or iron (as bis-glycinate).

The dietary focus on chemical elements derives from an interest in supporting the biochemical reactions of metabolism with the required elemental components.

Appropriate intake levels of certain chemical elements have been demonstrated to be required to maintain optimal health. Diet can meet all the body's chemical element requirements, although supplements can be used when some requirements (e.g., calcium, which is found mainly in dairy products) are not adequately met by the diet, or when chronic or acute deficiencies arise from pathology, injury, etc.

Research has supported that altering inorganic mineral compounds (carbonates, oxides, etc.) by reacting them with organic ligands (amino acids, organic acids, etc.) improves the bioavailability of the supplemented mineral.

Other Elements

Many elements have been suggested as essential, but such claims have usually not been confirmed. Definitive evidence for efficacy comes from the characterization of a biomolecule containing the element with an identifiable and testable function. One problem with identifying efficacy is that some elements are innocuous at low concentrations and are pervasive, so proof of efficacy is lacking because deficiencies are difficult to reproduce.

Mineral Ecology

Recent studies have shown a tight linkage between living organisms and minerals on this planet. This has let to the redefinition of minerals as "an element or compound, amorphous or crystalline, formed through 'biogeochemical' processes. The addition of `bio' reflects a greater appreciation, although an incomplete understanding, of the processes of mineral formation by living forms."

Biologists and geologists have only recently started to appreciate the magnitude of mineral biogeoengineering. Bacteria have contributed to the formation of minerals for billions of years and critically define the biogeochemical mineral cycles on this planet. Microorganisms can precipitate metals from solution contributing to the formation of ore deposits in addition to their ability to catalyze mineral dissolution, to respire, precipitate, and form minerals.

Most minerals are inorganic in nature. Mineral nutrients refers to the smaller class of minerals that are metabolized for growth, development, and vitality of living organisms. Mineral nutrients are recycled by bacteria that are freely suspended in the vast water columns of the worlds oceans.

They absorb dissolved organic matter containing mineral nutrients as they scavenge through the dying individuals that fall out of large phytoplankton blooms. Flagellates are effective bacteriovores and are also commonly found in the marine water column. The flagellates are preyed upon by zooplankton while the phytoplankton concentrates on the larger particulate matter that is suspended in the water column as they are consumed by larger zooplankton, with fish as the top predator. Mineral nutrients cycle through this marine food chain, from bacteria and phytoplankton to flagellates and zooplankton who are then eaten by fish. The bacteria are important in this chain because only they have the physiological ability to absorb the dissolved mineral nutrients from the sea. These recycling principals from marine environments apply to many soil and freshwater ecosystems as well.

THE RISKS OF BAD DIETS

The Potential Risks of Low-carbohydrate Diets: The American Heart Association, the American Dietetic Association and the American Kidney Fund warn about the potential risks from the long-term use of low-carbohydrate, high-protein diets. Low-carbohydrate diets, like Atkins diet avoid carbohydrates (starches or sugar) reducing or even eliminating the intake of fruit, fruit juices, starchy vegetables, beans, bread, rice, cereals pasta and other grain products. This leads to a nutrition consisting mainly of fat and proteins. Permitted are meat, cheese, non-starchy vegetables. Everything else is limited to a minimum. Later on the carbohydrate restriction is reduced but high fat and proteins persists. Despite initial weight loss higher than rob obtained with other diets, low-carb diets however have a net weight loss after one year which is not significantly different from other diets in comparison.

Many nutritionists attack the Atkins weight loss regime for its high fat content, increasing the risk of clogged arteries heart attack, diabetes, stroke and several types of cancer in the long term. Deleterious effects on cardiovascular disease risk factors were demonstrated by Larosa in a study of 24 obese individuals who followed the Atkins diet for 3 month. Caloric Intake declined, but LDL cholesterol levels rose despite the weight loss. Most of the weight loss occurred in the first few weeks, which suggests the combined effects of fluid loss and potential anorectic (reduction of desire to eat) effects of induced ketosis.

The American Heart Association does not recommend high-protein diets because they restrict foods that provide essential nutrients. An undersupply of vitamins and minerals, as well as potential cardiac, renal, bone and liver abnormalities may result.

A new study carried out at the University of Oxford demonstrates a direct affect on the heart caused by this diet. The results were presented at the American Heart Association conference in November 15, 2005, in Dalas. According to this study, the energy stores in the heart were reduced by an average of 16 per cent as a result of the Atkins diet recommendations. One of the authors, Dr. Damien Tyler says that similar patterns in a more severe form in patients with heart failure, and type 2 diabetics also suffer from low energy stores.

Evaluation of High-Protein Diets

Evaluation of high-protein diets must be done ensuring that It gives an additional recommendation not to eat simultaneously products rich in proteins together with dietary fibres.

Arguments against the diet of Dr. Atkins and the book "Fit for Live"

Traditional physiology does not agree with the diet of separation of Dr. Atkins. Meat does not have dietary fibres. It is therefore necessary to eat meat together with vegetables and fruits. The fibres absorb rests of digestion.

Protein digestion starts in the stomach with chloridric acid, together with pepsin. In the intestines there are proteases such as produced by the pancreas and by the gall bladder.

Important Proteasesare: Trypsin peptidase elastase ribonuclease and others.

As the external layer of the intestines is also composed by proteins the danger of proteases to digest his own meat. Nature protects against an own digestion secreting a protective layer of mucus made of mucopolysaccharides. This is the origin of the denomination"mucous membrane" which is not attacked by proteases. Sometimes however particles of enzymes get through the mucus barrier. damaging mucous cells which can regenerate rapidly.With protein rich nutrition there are more proteolytic enzymes secreted as in case which should of a normal nutrition.

Fibres of vegetables, fruits and cereals absorb excesses of enzymes.Fibres also accelerates passage of not digestible material through the intestines. The diet of Dr. Hay advises the ingestion of proteins at night when the body rests permitting the enzymes to stay longer in the intestines. The danger of intestinal cancer and other gastric malfunctions are possible.

Other Low-carbohydrate Diets

Zone Diet: The Zone Diet was developed by Barry Sears, a former researcher in biotechnology at the Massachusetts Institute of Technology. It is based on the control of insulin by balancing protein and carbohydrate at each meal (1-4). The American Heart Association notes that food must be eaten in required proportions of protein (40), fat (30), carbohydrates (30). Menues are not appealing. Vegetable portions are very large. Food is low in copper. Theories and long-term results are not validated. Couuld result in weight maintenance if carefully followed.

Protein Power: The Protein Power diet was developed by Michael R. Eades, M.D., and Mary Dan Eades, M.D. The principle of Protein Power is the restriction of carbohydrates-in particular refined ones. The American Heart Association notes that this diet is not practical for long term. Rigid rules must be observed. The diet is low in calcium, fibre, pantothenic acid, copper, manganese. It has limited food choice and is not practical for long term. It is high in total fat and saturated fat. No long term, validated studies are published.

Sugar Busters: Sugar busters involves eating high fibre vegetables and whole grains. Fruits are encouraged and discourages saturated fats and transfats. It is important to eat three regular meals daily. Moderation in portions size is most important. Exercise is also on the program. The American Heart Association notes that Sugar Buster eliminates many carbohydrate foods. Discourages eating fruit with meals. It is low in calcium, vitamin D, vitamin E, pantothenic acid, copper, potassium.

Stillman Diet: It is a severe form of ketogenic diet. Nothing is permitted if it is not mentioned in a list. The american Heart Association notes that Stillman diet is low in fibre, vitamin A, thiamine, vitamin C, vitamin D, folate, pantothenic acid, calcium, copper, magnesium, manganese, potassium. There are no long-term, valid studies published. There are extreme limitations in food choice, and very little variety.

Ketosis and the Ketogenic Diet: Elevated levels of ketones in the blood is called ketosis. It can occur during fasting, during a very high fat diet or during a long lasting deficiency in carbohydrate supply. This is the principle of the ketogenic diet. Atkins diet, which is used basically as low-glycemic index diet, may also cause ketosis. There are three ketones formed under hunger (also called " ketone bodies"): acetoacetate, acetone and beta-hydroxybutyrate (BHBA).

The fatty acids with participation of glucagon are digested in the liver to ketone compounds in order to liberate the needed energy. This is to see in opposition to the formation of D-glucose from carbohydrates to liberate energy. Ketones are burned in muscle and other tissues. High levels of ketones in tissue reduce the pH level. This is called ketoacidosis. One ketone is acetone which may cross the blood/brain barrier and, if in high concentration, causes central nervous disorders. Deficiency of insulin in diabetics may lead to ketosis.

It seems that fasting or high fat diet known as ketogenic diet is appropriate to induce ketosis for medical treatment of a specific neurological disorder. It should not be used as a body weight reducing diet.

Triglycerides used in Ketogenic Diets

Medium-chain triglyceride ketogenic diet for adults is being suggested by Schiff and colleagues. The researchers recommend a therapeutic trial of the ketogenic diet in intractable epilepsy for all ages. Foods used in the diet include high â€" triglyceride dairy products (e.g., butter, cream), mayonnaise and peanut butter. Carbohydrates, found in bread and starches, are eliminated in the diet, and liquid and calorie intake are often restricted as well in order to aid ketone accumulation.

Dr. Richard Veech at NIH in a study, concludes that ketone bodies increase metabolic efficiency, while decreasing production of free radicals, may treat neurological diseases such as Alzheimer's and Parkinson's disease, and the heart and brain operate 25% more efficiently using ketones as a source of energy. Alternative sources of ketones to avoid high fat diet for Alzheimer's and Parkinson's patients:

Richard Veech and colleagues (2000) found that the elevation of ketones may offer neuroprotection in the treatment or prevention of both Alzheimer's disease, where therapy is lacking, and Parkinson's disease, where therapy with L-dopa is time limited. The researchers recommend alternative sources of ketones, produced biotechnologically to overcome the atherogenic potential of the high-fat ketogenic diet used in childhood epilepsy.

The Mayo Diet

The Mayo diet has no connection with the famous Mayo-Clinic of USA. It is a protein rich and fat reduced diet (1.000 to 1.500 kcal) The diet includes 3 eggs every day, meat, fruits and vegetables. The diet is not well balanced. It is very high in cholesterol, and even for short time not suitable.

Ovo-lacto-vegetarian Diets

Bircher-Benner Diet: It is the ovo-lacto-vegetarian diet from Dr. Max Bircher-Benner which according to Dr. Bircher-Benner

- Mobilizes the ability of self healing
- Stimulates the intestines
- Fresh vegetable food containing high energy of the sun
- contains only three meals.

The Diet Contains Minimal Amount of Food: The diet contains lacto-vegetable food, at least 70% food of vegetable origin as muesli, fresh fruits, vegetable, salads, only moderate heat is used to prepare food. All ingredients should be of ecological origin.

No Potatoes and no Fish is Allowed: The Birchner-Benner diet avoids meat and meat products being therefore low in animal fatty acids, cholesterol and purine. This is positive in case of gout, coronary diseases and disorder of the fatty acid metabolism. Like every lacto-vegetarian diet undersupply of iodine and iron can take place.

Evers Diet from the Physician Josev Evers: The Evers Diet was at first developed to be applied in case of Multiple sclerosis. It is a moderate ovo-lacto-vegetarian diet with germinated cereals (containing high level of polyunsaturated fatty acids.) Ingredients should be of natural, unmodified origin. " Denaturated" food resulting from industrial processing was considered as being the reason for diseases of industrial era. The diet has different levels to be applied on different diseases:

- In case of disease only unheated vegetarian food, no potatoes are given.
- For convalescents low-fat meat, ground unheated meat, fresh fish and raw ham are given.

Comment

The ban of potatoes is unreasonable. Potatoes are low in calories, and dietary fibres and vitamin rich. Ground, unheated meat is a danger to weak patients because of possible bacterial hazard.

Waerland diet, Waerland, Sweden: This diet was established to clean intestines, to balance acid and base metabolism. Regulate the intestinal flora. Useful bacteria from milk and plants are in confrontation with harmful bacteria from meat, eggs and fish. The diet has low amount of food. Raw vegetable meals are followed by cereals meals. Milk, fruits and vegetables are considered as alkaline food. Instructions about sleeping and working time are given. In the morning 1/2 litre of potato-vegetable soup is taken followed by warm bed rest lying on stomach. The Waerland diet cannot be explained. Meat, eggs and fish cannot be considered as harmful and many claims are not true.

Danger of Sudden Changes of Nutrition

Increasing Salads and other vegetables increase intake of certain vitamins like folate, vitamin A and K. Decreasing high caloric foods like dairy products, red meat, salad oil or eggs reduces intake of minerals such as calcium zinc and selenium, vitamin D, E, B12 and essential fatty acids.

Anthroposophical Nutrition

It was established by Rudolf Steiner. It should develop consciousness, to live in reconciliation with nature, nutrition as part of the holistic doctrine. Ovo-lacto-vegetable nutrition with low content of meat. Avoid potatoes as they destroy instinct and promote materialism. Ban of use of pressure cooking and frozen food as they are "enemies of life".

Comment

Many of the doctrines of anthroposophy cannot be followed. The ban of potatoes, pressure cooking and frozen food is not tenable.

Macrobiotic Nutrition According M. Kushi

Macrobiotic nutrition was founded by G.Ohsawa, Japan. His nutrition was dangerous and some people died using this diet. In USA macrobiotic nutrition according Ohsawa was prohibited. M. Kushi modernized the content of macrobiotic nutrition. A short use of this nutrition is harmless. It is harmful and not indicated to be followed for a log period. Macrobiotic nutrition is a part of Zen-Buddhism a kind of philosophy of life. Life energy comes from vegetable nutrition, cereals in the highest form. Contrast from passive to active, from night to day, from feminine to masculine, is the principle of Yin and Yang which were transferred to food by Kushi. No milk, and milk products, no meat and tea are consumed.

The Jo-Jo Effect Reviewed

Two important studies recently reviewed the knowledge about the Jo-Jo effect, following a diet. These studies received much attention in 2007 following the effort to tackle the obesity problem in large parts of the population of developed countries. Rena Wing and James Hill of the University of Colorado started in 1994 " The National Weight-Control Register". In Germany Joachim Westenhöfer of the Hamburger Hochschule für Angewandte Wissenschaften started years ago the "Lean Habits Study".

The Strategy: Both studies agreed with foregoing studies in the strategy to counter overweight:

- Reduction of intake of energy.
- Increase of energy
- Overcome of the own psyche
- Only one of these three items may become a great help to reduce weight and keep it down, but using a bit of all three recommendations has the best chance to avoid Jo-Jo effect.

The Physiology of Overweight and Jo-Jo Effect

The authors stress the fact that a variety of reasons account for overweight and a jo-jo effect following a diet. The body reacts to weight loss due to famine or diet with defence mechanisms such as increased hunger and reduced energy consume.This makes weight-reduction and to keep it down so difficult

- *Thyroid Gland:* The thyroid glad reduces the production of hormones as soon as the body weight diminishes. Some parts of the nervous system are also less active reducing thus the metabolism.
  Less calories are transformed.
- *Adipous Tissue:* The adipous tissue elaborates leptin, a compound which signalises the brain to increase the feeling of hunger, and the body temperature is slightly reduced, reducing thus the basal metabolism.

- *Stomach:* The stomach produces ghrelin which is an hormone triggering hunger feeling in the brain.
- *Pancreases:* The pancreases produces insulin which is reduced when the body weight falls. Less insulin brings the brain to increase appetite and reduces energy consumption.
- *Muscle Mass:* Body weight reduction is related with loss of muscle mass, this means that the overall consumption of calories is also reduced resulting again in weight gain.

The studies clearly state that overweight is not bound to a genetic code, what is settled is the individual energy output. Seven strategies are recommended:

1. *Stay Mobile:* Sport and mobility during work reduces the risk of overweight. The main goal of sport and mobility is to keep, or build muscle mass. Walking, swimming or biking are recommended by the authors.
    - Make sport of 30 minutes three times a week
    - Don't use elevators. Use stairs instead of escalators. Use your bike or walk to your office.
2. *Never Say "Never":* Trying to avoid everything brings frustration the moment you make a small error. You may have a nice fatty dinner as long you compensate it the next day with vegetables or you insert an extra sport session.
3. *Don't Loose the Rhythm:* Eat always at the same time. Take your time for your meals.do not eat in a hurry. Do not miss your breakfast.
4. *Controll is Necessary:* Weigh yourself once a day or at least once a week, so you can take countermeasures as soon as your weight rises.
5. *Be Stingy with Fat:* You can do without butter, fatty cheese, heavy sauces and fatt pork. Instead of this you can eat vegetables and fruits instead.
6. *Fresh Vegetables:* You can eat as much vegetables and salad. as you wish. You get enough dietary fibre and have a low calorie intake. Eat small portions of fruits spare with banana, raisins, grapes and dates.
7. *Join a Weight Club:* You may feel strong in company of kindred spirit in a weight club. Avoid people who seduce you to eat the wrong foods.

NUTRITION

Foods high in carbohydrates include breads, pastas, beans, potatoes, bran, rice, and cereals. Most such foods are high in starch. Carbohydrates are the most common source of energy in living organisms. Proteins and fat are necessary building components for body tissue and cells, and are also a source of energy for most organisms. Carbohydrates are not essential nutrients in humans: the body can obtain all its energy from protein and fats. The brain and neurons generally cannot burn fat for energy, but can use glucose or ketones; the body can also synthesize some glucose from a few of the amino acids in protein and also from the glycerol backbone in triglycerides.

Carbohydrate contains 15.8 kilojoules (3.75 kilocalories) and proteins 16.8 kilojoules (4 kilocalories) per gram, while fats contain 37.8 kilojoules (9 kilocalories) per gram.

In the case of protein, this is somewhat misleading as only some amino acids are usable for fuel. Likewise, in humans, only some carbohydrates are usable for fuel, as in many monosaccharides and some disaccharides. Other carbohydrate types can be used, but only with the assistance of gut bacteria. Ruminants and termites can even process cellulose, which is indigestible to humans.

Based on the effects on risk of heart disease and obesity, the Institute of Medicine recommends that American and Canadian adults get between 45–65% of dietary energy from carbohydrates. The Food and Agriculture Organization and World Health Organization jointly recommend that national dietary guidelines set a goal of 55–75% of total energy from carbohydrates, but only 10% directly from sugars (their term for simple carbohydrates).

Classification

For dietary purposes, carbohydrates can be classified as simple (monosaccharides and disaccharides) or complex (oligosaccharides and polysaccharides). The term *complex carbohydrate* was first used in the U.S. Senate Select Committee on Nutrition and Human Needs publication *Dietary Goals for the United States* (1977), where it denoted "fruit, vegetables and whole-grains". Dietary guidelines generally recommend that complex carbohydrates, and such nutrient-rich simple carbohydrate sources such as fruit (glucose or fructose) and dairy products (lactose) make up the bulk of carbohydrate consumption. This excludes such sources of simple sugars as candy and sugary drinks. The USDA's *Dietary Guidelines for Americans 2005* dispensed with the simple/complex distinction, instead recommending fibre-rich foods and whole grains.

The glycemic index and glycemic load concepts have been developed to characterize food behaviour during human digestion. They rank carbohydrate-rich foods based on the rapidity of their effect on blood glucose levels. The insulin index is a similar, more recent classification method that ranks foods based on their effects on blood insulin levels, which are caused by glucose (or starch) and some amino acids in food. Glycemic index is a measure of how quickly food glucose is absorbed, while glycemic load is a measure of the total absorbable glucose in foods.

CARBOHYDRATE

A carbohydrate is an organic compound with the empirical formula $C_m(H_2O)_n$ (where *m* could be different from *n*); that is, consists only of carbon, hydrogen, and oxygen, with a hydrogen:oxygen atom ratio of 2:1 (as in water). However, there are exceptions to this. One common example would be deoxyribose, a component of DNA, which has the empirical formula $C_5H_{10}O_4$. Carbohydrates can be viewed as hydrates of carbon, hence their name. Structurally however, it is more accurate to view them as polyhydroxy aldehydes and ketones. The term is most common in biochemistry, where it is a synonym of saccharide. The carbohydrates (saccharides) are divided into four chemical groupings: monosaccharides, disaccharides, oligosaccharides, and polysaccharides. In general, the monosaccharides and disaccharides, which are smaller (lower molecular weight) carbohydrates, are commonly referred to as sugars. The word *saccharide* comes from the Greek word *sákkharon*, meaning "sugar". While the scientific nomenclature of carbohydrates is complex, the names of the monosaccharides and disaccharides very often end in the suffix-ose. For example, blood sugar is the monosaccharide glucose, table sugar is the disaccharide sucrose, and milk sugar is the disaccharide lactose.

Carbohydrates perform numerous roles in living things. Polysaccharides serve for the storage of energy (e.g., starch and glycogen), and as structural components (e.g., cellulose in plants and chitin in arthropods). The 5-carbon monosaccharide ribose is an important component of coenzymes (e.g., ATP, FAD, and NAD) and the backbone of the genetic molecule known as RNA. The related deoxyribose is a component of DNA. Saccharides and their derivatives include many other important biomolecules that play key roles in the immune system, fertilization, preventing pathogenesis, blood clotting, and development.

In food science and in many informal contexts, the term carbohydrate often means any food that is particularly rich in the complex carbohydrate starch (such as cereals, bread, and pasta) or simple carbohydrates, such as sugar (found in candy, jams, and desserts).

Structure

Formerly the name "carbohydrate" was used in chemistry for any compound with the formula $C_m(H_2O)_n$. Following this definition, some chemists considered formaldehyde (CH_2O) to be the simplest carbohydrate, while others claimed that title for glycolaldehyde. Today the term is generally understood in the biochemistry sense, which excludes compounds with only one or two carbons.

Natural saccharides are generally built of simple carbohydrates called monosaccharides with general formula $(CH_2O)_n$ where *n* is three or more. A typical monosaccharide has the structure H-$(CHOH)_x$(C=O)-$(CHOH)_y$-H, that is, an aldehyde or ketone with many hydroxyl groups added, usually one on each carbon atom that is not part of the aldehyde or ketone functional group. Examples of monosaccharides are glucose, fructose, and glyceraldehyde. However, some biological substances commonly called "monosaccharides" do not conform to this formula (e.g., uronic acids and deoxy-sugars such as fucose), and there are many chemicals that do conform to this formula but are not considered to be monosaccharides (e.g., formaldehyde CH_2O and inositol $(CH_2O)_6$).

The open-chain form of a monosaccharide often coexists with a closed ring form where the aldehyde/ketone carbonyl group carbon (C=O) and hydroxyl group (-OH) react forming a hemiacetal with a new C-O-C bridge.

Monosaccharides can be linked together into what are called polysaccharides (or oligosaccharides) in a large variety of ways. Many carbohydrates contain one or more modified monosaccharide units that have had one or more groups replaced or removed. For example, deoxyribose, a component of DNA, is a modified version of ribose; chitin is composed of repeating units of N-acetylglucosamine, a nitrogen-containing form of glucose.

Monosaccharides

Monosaccharides are the simplest carbohydrates in that they cannot be hydrolyzed to smaller carbohydrates. They are aldehydes or ketones with two or more hydroxyl groups. The general chemical formula of an unmodified monosaccharide is $(CH_2O)_n$, literally a "carbon hydrate." Monosaccharides are important fuel molecules as well as building blocks for nucleic acids. The smallest monosaccharides, for which n = 3, are dihydroxyacetone and D-and L-glyceraldehyde.

Classification of Monosaccharides

Monosaccharides are classified according to three different characteristics: the placement of its carbonyl group, the number of carbon atoms it contains, and its chiral handedness. If the carbonyl group is an aldehyde, the monosaccharide is an aldose; if the carbonyl group is a ketone, the monosaccharide is a ketose. Monosaccharides with three carbon atoms are called trioses, those with four are called tetroses, five are called pentoses, six are hexoses, and so on. These two systems of classification are often combined.

For example, glucose is an aldohexose (a six-carbon aldehyde), ribose is an aldopentose (a five-carbon aldehyde), and fructose is a ketohexose (a six-carbon ketone). Each carbon atom bearing a hydroxyl group (-OH), with the exception of the first and last carbons, are asymmetric, making them stereocenters with two possible configurations each (R or S). Because of this asymmetry, a number of isomers may exist for any given monosaccharide formula.

The aldohexose D-glucose, for example, has the formula $(CH_2O)_6$, of which all but two of its six carbons atoms are stereogenic, making D-glucose one of $2^4 = 16$ possible stereoisomers. In the case of glyceraldehyde, an aldotriose, there is one pair of possible stereoisomers, which are enantiomers and epimers. 1,3-dihydroxyacetone, the ketose corresponding to the aldose glyceraldehyde, is a symmetric molecule with no stereocenters). The assignment of D or L is made according to the orientation of the asymmetric carbon furthest from the carbonyl group: in a standard Fischer projection if the hydroxyl group is on the right the molecule is a D sugar, otherwise it is an L sugar. The "D-" and "L-" prefixes should not be confused with "d-" or "l-", which indicate the direction that the sugar rotates plane polarized light. This usage of "d-" and "l-" is no longer followed in carbohydrate chemistry.

Ring-straight Chain Isomerism

The aldehyde or ketone group of a straight-chain monosaccharide will react reversibly with a hydroxyl group on a different carbon atom to form a hemiacetal or hemiketal, forming a heterocyclic ring with an oxygen bridge between two carbon atoms. Rings with five and six atoms are called furanose and pyranose forms, respectively, and exist in equilibrium with the straight-chain form.

During the conversion from straight-chain form to the cyclic form, the carbon atom containing the carbonyl oxygen, called the anomeric carbon, becomes a stereogenic centre with two possible configurations: The oxygen atom may take a position either above or below the plane of the ring. The resulting possible pair of stereoisomers are called anomers. In the *α anomer*, the-OH substituent on the anomeric carbon rests on the opposite side (trans) of the ring from the CH_2OH side branch. The alternative form, in which the CH_2OH substituent and the anomeric hydroxyl are on the same side (cis) of the plane of the ring, is called the *β anomer*. You can remember that the β anomer is cis by the mnemonic, "It's always better to βe up". Because the ring and straight-chain forms readily interconvert, both anomers exist in equilibrium. In a Fischer Projection, the α anomer is represented with the anomeric hydroxyl group *trans* to the CH_2OH and *cis* in the β anomer.

Use in Living Organisms

Monosaccharides are the major source of fuel for metabolism, being used both as an energy source (glucose being the most important in nature) and in biosynthesis. When monosaccharides are not immediately needed by many cells they are often converted to more space efficient forms, often polysaccharides. In many animals, including humans, this storage form is glycogen, especially in liver and muscle cells. In plants, starch is used for the same purpose.

Disaccharides

Two joined monosaccharides are called a disaccharide and these are the simplest polysaccharides. Examples include sucrose and lactose. They are composed of two monosaccharide units bound together by a covalent bond known as a glycosidic linkage formed via a dehydration reaction, resulting in the loss of a hydrogen atom from one monosaccharide and a hydroxyl group from the other. The formula of unmodified disaccharides is $C_{12}H_{22}O_{11}$. Although there are numerous kinds of disaccharides, a handful of disaccharides are particularly notable.

Sucrose, pictured to the right, is the most abundant disaccharide, and the main form in which carbohydrates are transported in plants. It is composed of one D-glucose molecule and one D-fructose molecule. The systematic name for sucrose, *O*-α-D-glucopyranosyl-(1→2)-D-fructofuranoside, indicates four things:

- Its monosaccharides: glucose and fructose
- Their ring types: glucose is a pyranose, and fructose is a furanose
- How they are linked together: the oxygen on carbon number 1 (C1) of α-D-glucose is linked to the C2 of D-fructose.
- The-*oside* suffix indicates that the anomeric carbon of both monosaccharides participates in the glycosidic bond.

Lactose, a disaccharide composed of one D-galactose molecule and one D-glucose molecule, occurs naturally in mammalian milk. The systematic name for lactose is *O*-β-D-galactopyranosyl-(1→4)-D-glucopyranose. Other notable disaccharides include maltose (two D-glucoses linked α-1,4) and cellulobiose (two D-glucoses linked β-1,4). disaccharides can be classified into two types. They are reducing and non-reducing disaccahrides if the functional group is present in bonding with another sugar unit it is called as reducing disaccharide.

Oligosaccharides and Polysaccharides

Oligosaccharides and polysaccharides are composed of longer chains of monosaccharide units bound together by glycosidic bonds. The distinction between the two is based upon the number of monosaccharide units present in the chain. Oligosaccharides typically contain between three and ten monosaccharide units, and polysaccharides contain greater than ten monosaccharide units. Definitions of how large a carbohydrate must be to fall into each category vary according to personal opinion. Examples of oligosaccharides include the disaccharides mentioned above, the trisaccharide raffinose and the tetrasaccharide stachyose.

Oligosaccharides are found as a common form of protein posttranslational modification. Such posttranslational modifications include the Lewis and ABO oligosaccharides responsible for blood group classifications and so of tissue incompatibilities, the alpha-Gal epitope responsible for hyperacute rejection in xenotransplantation, and O-GlcNAc modifications.

Polysaccharides represent an important class of biological polymers. Their function in living organisms is usually either structure-or storage-related. Starch (a polymer of glucose) is used as a storage polysaccharide in plants, being found in the form of both amylose and the branched amylopectin. In animals, the structurally similar glucose polymer is the more densely branched glycogen, sometimes called 'animal starch'. Glycogen's properties allow it to be metabolized more quickly, which suits the active lives of moving animals.

Cellulose and chitin are examples of structural polysaccharides. Cellulose is used in the cell walls of plants and other organisms, and is claimed to be the most abundant organic molecule on earth. It has many uses such as a significant role in the paper and textile industries, and is used as a feedstock for the production of rayon (via the viscose process), cellulose acetate, celluloid, and nitrocellulose. Chitin has a similar structure, but has nitrogen-containing side branches, increasing its strength. It is found in arthropod exoskeletons and in the cell walls of some fungi. It also has multiple uses, including surgical threads.

Other polysaccharides include callose or laminarin, chrysolaminarin, xylan, arabinoxylan, mannan, fucoidan and galactomannan.

Nutrition

Foods high in carbohydrate include fruits, sweets, soft drinks, breads, pastas, beans, potatoes, bran, rice, and cereals. Carbohydrates are a common source of energy in living organisms; however, no carbohydrate is an essential nutrient in humans.

Carbohydrates are not necessary building blocks of other molecules, and the body can obtain all its energy from protein and fats. The brain and neurons generally cannot burn fat for energy, but use glucose or ketones. Humans can synthesize some glucose (in a set of processes known as gluconeogenesis) from specific amino acids, from the glycerol backbone in triglycerides and in some cases from fatty acids. Carbohydrate contains 15.8 kilojoules (3.75 kilocalories) and proteins 16.8 kilojoules (4 kilocalories) per gram, while fats contain 37.8 kilojoules (9 kilocalories) per gram. In the case of protein, this is somewhat misleading as only some amino acids are usable for fuel. Organisms typically cannot metabolize all types of carbohydrate to yield energy. Glucose is a nearly universal and accessible source of calories. Many organisms also have the ability to metabolize other monosaccharides and Disaccharides, though glucose is preferred. In *Escherichia coli*, for example, the *lac* operon will express enzymes for the digestion of lactose when it is present, but if both lactose and glucose are present the *lac* operon is repressed, resulting in the glucose being used first. Polysaccharides are also common sources of energy. Many organisms can easily break down starches into glucose, however, most organisms cannot metabolize cellulose or other polysaccharides like chitin and arabinoxylans. These carbohydrates types can be metabolized by some bacteria and protists. Ruminants and termites, for example, use microorganisms to process cellulose. Even though these complex carbohydrates are not very digestible, they may comprise important dietary elements for humans. Called dietary fiber, these carbohydrates enhance digestion among other benefits.

Based on the effects on risk of heart disease and obesity, the Institute of Medicine recommends that American and Canadian adults get between 45–65% of dietary energy from carbohydrates. The Food and Agriculture Organization and World Health Organization jointly recommend that national dietary guidelines set a goal of 55–75% of total energy from carbohydrates, but only 10% directly from sugars (their term for simple carbohydrates).

Classification

Historically nutritionists have classified carbohydrates as either simple or complex, however, the exact delineation of these categories is ambiguous. Today, simple carbohydrate typically refers to monosaccharides and disaccharides and complex carbohydrate means polysaccharides (and oligosaccharides). However, the term *complex carbohydrate* was first used in slightly different context in the U.S. Senate Select Committee on Nutrition and Human Needs publication *Dietary Goals for the United States* (1977). In this work, complex carbohydrate were defined as "fruit, vegetables and whole-grains". Some nutritionists use complex carbohydrate to refer to any sort of digestible saccharide present in a whole food, where fiber, vitamins and minerals are also found (as opposed to processed carbohydrates, which provide calories but few other nutrients).

A commonly held belief, even among nutritionists, is that complex carbohydrates (polysaccharides, e.g. starches) are digested more slowly than simple carbohydrates (sugars) and thus are healthier. However, there appears to be no significant difference between simple and complex carbohydrates in terms of their effect on blood sugar. Some simple carbohydrates (e.g. fructose) are digested very slowly, while some complex carbohydrates (starches), especially if processed, raise blood sugar rapidly. The speed of digestion is determined by a variety of factors including which other nutrients are consumed with the carbohydrate, how the food is prepared, individual differences in metabolism, and the chemistry of the carbohydrate.

The glycemic index (GI) and glycemic load concepts have been developed to characterize food behavior during human digestion. They rank carbohydrate-rich foods based on the rapidity and magnitude of their effect on blood glucose levels. Glycemic index is a measure of how quickly food glucose is absorbed, while glycemic load is a measure of the total absorbable glucose in foods. The insulin index is a similar, more recent classification method that ranks foods based on their effects on blood insulin levels, which are caused by glucose (or starch) and some amino acids in food.

Dietary guidelines generally recommend that complex carbohydrates (starches) and nutrient-rich simple carbohydrates such as fruits and vegetables, and dairy products make up the bulk of carbohydrate consumption. Highly processed sources of carbohydrate such as corn or potato chips, candy, sugary drinks, pastries and white rice are generally considered unhealthy in excess. The USDA's *Dietary Guidelines for Americans 2005* dispensed with the simple/complex distinction, instead recommending fiber-rich foods and whole grains.

Metabolism—Catabolism

Catabolism is the metabolic reaction cells undergo to extract energy. There are two major metabolic pathways of monosaccharide catabolism: glycolysis and the citric acid cycle.

In glycolysis, oligo/polysaccharides are cleaved first to smaller monosaccharides by enzymes called glycoside hydrolases. The monosaccharide units can then enter into monosaccharide catabolism. In some cases, as with humans, not all carbohydrate types are usable as the digestive and metabolic enzymes necessary are not present.

Carbohydrate Chemistry

Carbohydrate chemistry is a large and economically important branch of organic chemistry. Some of the main organic reactions that involve carbohydrates are:

- Carbohydrate acetalisation
- Cyanohydrin reaction
- Lobry-de Bruyn-van Ekenstein transformation

- Amadori rearrangement
- Nef reaction
- Wohl degradation
- Koenigs–Knorr reaction.

PROTEIN (NUTRIENT)

Proteins are polymer chains made of amino acids linked together by peptide bonds. Amino acids can be divided into either essential amino acids or non-essential amino acids. Proteins and carbohydrates contain 4 kcal per gram as opposed to lipids which contain 9 kcal per gram. The essential amino acids, which must be obtained from food sources, are leucine, isoleucine, valine, lysine, threonine, tryptophan, methionine, phenylalanine and histidine. On the other hand, non-essential amino acids can be made by the body from other amino acids. The non-essential amino acids are arginine, alanine, asparagine, aspartic acid, cysteine, glutamine, glutamic acid, glycine, proline, serine, and tyrosine.

In nutrition, proteins are broken down in the stomach during digestion by enzymes known as proteases into smaller polypeptides to provide amino acids for the body, including the essential amino acids that cannot be biosynthesized by the body itself. Thus, protein from one's diet should provide both essential and non-essential amino acids for protein synthesis. Most animal and certain vegetable proteins are considered complete proteins with a full complement of essential amino acids in adequate proportions. People who avoid animal products may practice protein combining to get the essential amino acids in their diet.

Protein Functions in Body

Protein is a nutrient needed by the human body for growth and maintenance. Aside from water, protein is the most abundant molecule in the body. Protein is found in all cells of the body and is the major structural component of all cells in the body, especially muscle. This also includes body organs, hair and skin. Proteins also are utilized in membranes, such as glycoproteins. When broken down into amino acids, they are used as precursors to nucleic acid and vitamins. Hormones and enzymes are also formed from amino acids in which they help regulate metabolism, support the immune system and other body functions. Finally, protein is needed to form blood cells.

Protein Function in Exercise

Proteins are one of the key nutrients for success in terms of sports. They play a major role in the response to exercise. Amino acids, the building blocks of proteins, are used for building new tissue, including muscle, as well as repairing damaged tissues. Proteins, however, only provide a small source of fuel for the exercising muscles, being used as fuel typically only when carbohydrates and lipid resources are low.

Sources

There are many different sources of protein ranging from whole protein foods (such as milk, meat, fish, egg, and vegetables) to a variety of protein powders (such as casein, whey, egg, rice and soy). Protein powders are processed and manufactured sources of protein. Protein powders may provide an additional source of protein for exercising muscles. The type of protein is important in terms of its influence on protein metabolic response and possibly on the muscle's exercise performance. The different physical and/or chemical properties within the various types of protein may affect the rate of protein digestion. As a result, the amino acid availability and the accumulation of tissue protein is altered because of the various protein metabolic responses.

Food	*Amount of protein (grams)*
Spirulina 1 cup	64
Tempeh 1 cup	41
Dried Parsley 1 cup	31
Lentils, cooked 1 cup	18
Black Beans, cooked 1 cup	15
Tofu, firm 4 oz	11
Quinoa, cooked 1 cup	9
Peanut Butter 2 tbsp	8
Almonds 1/4 cup	8
Sun-dried Tomato 1 cup	8
Brown rice, cooked 1 cup	5
Broccoli, cooked 1 cup	4
Potato 1 med.	4
Lambsquarters 1 cup	4

This table compares the different plant foods with the amount of protein in grams per serving

Protein Quality

Different proteins have different levels of biological availability (BA) to the human body. Many methods have been introduced to measure protein utilization and retention rates in humans. They include biological value, net protein utilization, and PDCAAS (Protein Digestibility Corrected Amino Acids Score) which was developed by the FDA as an improvement over the Protein

Efficiency Ratio (PER) method. These methods examine which proteins are most efficiently used by the body. The PDCAAS rating is a fairly recent evaluation method; it was adopted by the US Food and Drug Administration (FDA) and the Food and Agricultural Organization of the United Nations/ World Health Organization (FAO/WHO) in 1993 as "the preferred 'best'" method to determine protein quality. These organizations have suggested that other methods for evaluating the quality of protein are inferior.

Digestion

Most proteins are decomposed to single amino acids in digestion.

Digestion typically begins in the stomach when pepsinogen is converted to pepsin by the action of hydrochloric acid, and continued by trypsin and chymotrypsin in the intestine. Before the absorption in the small intestine, most proteins are already reduced to single amino acid or peptides of several amino acids. Most of peptides longer than 4 amino acids are not absorbed. Absorption into the intestinal absorptive cells is not the end. There most of peptides are broken into single amino acids.

Absorption of the amino acids and their derivatives into which dietary protein is degraded is done by the gastrointestinal tract.

The absorption rates of individual amino acids are highly dependent on the protein source; for example, the digestibilities of many amino acids in humans, the difference between soy and milk proteins and between individual milk proteins, beta-lactoglobulin and casein. For milk proteins, about 50% of the ingested protein is absorbed between the stomach and the jejunum and 90% is absorbed by the time the digested food reaches the ileum. Biological value (BV) is a measure of the proportion of absorbed protein from a food which becomes incorporated into the proteins of the organism's body.

Newborn

Newborns of mammals are exceptional in protein digestion and assimilation in that they can absorb intact proteins at the small intestine. This enables passive immunity from milk.

Dietary Requirements

Considerable debate has taken place regarding issues surrounding protein intake requirements. How much protein is needed in a person's daily diet, is determined in large part by overall energy intake, by the body's need for nitrogen and essential amino acids, body weight and composition, rate of growth in the individual, physical activity level, individual's energy and carbohydrate intake, as well as the presence of illness or injury. Physical activity and exertion as well as enhanced muscular mass increase the need for protein. Requirements are also greater during childhood for growth and development, during pregnancy or when breast-feeding in order to nourish a baby, or when the body needs to recover from malnutrition or trauma or after an operation.

If enough energy is not taken in through diet, as in the process of starvation, the body will use protein from the muscle mass to meet its energy needs, leading to muscle wasting over time. If the individual does not consume adequate protein in nutrition, then muscle will also waste as more vital cellular processes (e.g. respiration enzymes, blood cells) recycle muscle protein for their own requirements. According to US & Canadian Dietary Reference Intake guidelines, women aged 19–70 need to consume 46 grams of protein per day, while men aged 19–70 need to consume 56 grams of protein per day to avoid a deficiency. The American and Canadian guidelines recommend a daily protein dietary allowance, measured as intake per kilogram body weight, is 0.8 g/kg. However, this recommendation is based on structural requirements, but disregards use of protein for energy metabolism. This requirement is for a normal sedentary person.

Several studies have concluded that active people and athletes may require elevated protein intake (compared to 0.8 g/kg) due to increase in muscle mass and sweat losses, as well as need for body repair and energy source. Suggested amounts vary between 1.6 g/kg and 1.8 g/kg, while a proposed *maximum* daily protein intake would be approximately 25% of energy requirements i.e. approximately 2 to 2.5 g/kg. However, many questions still remain to be resolved.

Aerobic Exercise Protein Needs

Endurance athletes differ from strength-building athletes in that endurance athletes do not build muscle mass from training. Research suggests that individuals performing endurance activity require more protein intake than sedentary individuals so that muscles broken down during endurance workouts can be repaired. Although the protein requirement for athletes still remains controversial, research does show that endurance athletes can benefit from increasing protein intake because the type of exercise endurance athletes participate in still alters the protein metabolism pathway. The overall protein requirement increases because of amino acid oxidation in endurance-trained athletes. Endurance athletes who exercise over a long period (2–5 hours per training session) use protein as a source of 5–10% of their total energy expended. Therefore, a slight increase in protein intake may be beneficial to endurance athletes by replacing the protein lost in energy expenditure and protein lost in repairing muscles. Some scientists suggest that endurance athletes may increase daily protein intake to a maximum of 1.2–1.4 g per kg body weight.

Anaerobic Exercise Protein Needs

Research also indicates that individuals performing strength-training activity require more protein than sedentary individuals. Strength-training athletes may increase their daily protein intake to a maximum of 1.4–1.8 g per kg body weight to enhance muscle protein synthesis, or to make up for the loss of amino acid oxidation during exercise. Many athletes maintain a high-protein diet as part of their training, and so protein deficiency is less likely among this group than among non-athletes. In fact, some athletes who specialize in anaerobic sports (e.g. weightlifting) assume a very high level of protein intake is necessary, and may over-consume. Research indicates that many athletes consume more protein than they need even without the use of protein supplements.

Excess Consumption

When a high dietary protein intake is consumed, there is an increase in urea excretion, which suggests that amino acid oxidation is increased. High levels of protein intake increase the activity of branched-chain ketoacid dehydrogenase. As a result, oxidation is facilitated, and the amino group of the amino acid is excreted to the liver. This process suggests that excess protein consumption results in protein oxidation and that the protein is excreted. The body is unable to store excess protein. Protein is digested into amino acids, which enter the bloodstream. Excess amino acids are converted to other usable molecules by the liver in a process called deamination. Deamination converts nitrogen from the amino acid into ammonia, which is converted by the liver into urea in the urea cycle. Excretion of urea is performed by the kidneys. These organs can normally cope with any extra workload, but, if a kidney disease occurs, a decrease in protein will often be prescribed. Furthermore, as noted, protein provides the body with 4 calories per gram, and when there is excess protein intake, the body will utilize as much of it for energy as possible. After that stage, the body will produce fat from the excess protein, turning it into fat cells. On the other hand, if people do not eat enough calories, body protein and protein from the food will be utilized for energy. This is not ideal as the main function of protein is to maintain the body. Finally, when food protein intake is periodically high or low, the body tries to keep protein levels at an equilibrium by using the "labile protein reserve", which serves as a short-term protein store to be used for emergencies or daily variations in protein intake. However, that reserve is not utilized as longer-term storage for future needs.

Many researchers have also found that excessive intake of protein increases calcium excretion in urine. It has been thought that this occurs to maintain the pH imbalance from the oxidation of sulfur amino acids. Also, it is inconclusive whether bone resorption contributes to bone loss and osteoporosis. However, it is also found that a regular intake of calcium would be able to stabilize this loss.

Another issue arising from overconsumption of protein is a higher risk of kidney stone formation from calcium in the renal circulatory system. It has been found that high animal protein intake in healthy individuals increases the probability of forming kidney stones by 250 percent.

Testing in Foods

The classic assays for protein concentration in food are the Kjeldahl method and the Dumas method. These tests determine the total nitrogen in a sample. The only major component of most food which contains nitrogen is protein (fat, carbohydrate and dietary fibre do not contain nitrogen). If the amount of nitrogen is multiplied by a factor depending on the kinds of protein expected in the food the total protein can be determined. This value is known as the "crude protein" content. On food labels the protein is given by the nitrogen multiplied by 6.25, because the average nitrogen content of proteins is about 16%. The Kjeldahl test is typically used because it is the method the AOAC International has adopted and is therefore used by many food standards agencies around the world, though the Dumas method is also approved by some standards organizations.

Accidental contamination and intentional adulteration of protein meals with non-protein nitrogen sources that inflate crude protein content measurements have been known to occur in the food industry for decades. To ensure food quality, purchasers of protein meals routinely conduct quality control tests designed to detect the most common non-protein nitrogen contaminants, such as urea and ammonium nitrate.

In at least one segment of the food industry, the dairy industry, some countries (at least the U.S., Australia, France and Hungary), have adopted "true protein" measurement, as opposed to crude protein measurement, as the standard for payment and testing: "True protein is a measure of only the proteins in milk, whereas crude protein is a measure of all sources of nitrogen and includes nonprotein nitrogen, such as urea, which has no food value to humans.... Current milk-testing equipment measures peptide bonds, a direct measure of true protein." Measuring peptide bonds in grains has also been put into practice in several countries including Canada, the UK, Australia, Russia and Argentina where near-infrared reflectance (NIR) technology, a type of infrared spectroscopy is used. The Food and Agriculture Organization of the United Nations (FAO) recommends that only amino acid analysis be used to determine protein in, *inter alia*, foods used as the sole source of nourishment, such as infant formula, but also provides: "When data on amino acids analyses are not available, determination of protein based on total N content by Kjeldahl (AOAC, 2000) or similar method... is considered acceptable." The limitations of the Kjeldahl method were at the heart of the Chinese protein export contamination in 2007 and the 2008 Chinese Milk Scandal in which the industrial chemical melamine was added to the milk or glutens to increase the measured "protein".

Food Allergies

Specific proteins found in certain food items are often the cause of allergies and allergic reactions. This is because the structure of each form of protein is slightly different; some may trigger a response from the immune system while others remain harmless. Many people are allergic to casein, the protein in milk; gluten, the protein in wheat and other grains; the particular proteins found in peanuts; or those in shellfish or other seafoods. Food allergies should not be confused with food intolerance.

Deficiency in Developing Countries

Protein deficiency is a serious cause of ill health and death in developing countries. Protein deficiency plays a part in the disease kwashiorkor. Famine, overpopulation and other factors can increase rates of malnutrition and protein deficiency. Protein deficiency can lead to reduced intelligence or mental retardation.

In countries that suffer from widespread protein deficiency, food is generally full of plant fibers, which makes adequate energy and protein consumption very difficult. Protein deficiency is generally caused by lack of total food energy, making it an issue of not getting food in total. Symptoms of kwashiorkor include apathy, diarrhea, inactivity, failure to grow, flaky skin, fatty liver, and edema of the belly and legs. This edema is explained by the normal functioning of proteins in fluid balance and lipoprotein transport.

Moringa trees are known to overcome protein deficiency in developing countries as the leaves and other parts of the tree contain comparably to soy bean high amount of crude proteins and amino acids.

Dr. Latham, director of the Program in International Nutrition at Cornell University claims that malnutrition is a frequent cause of death and disease in third world countries. Protein-energy malnutrition (PEM) affects 500 million people and kills 10 million annually. In severe cases white blood cell numbers decline and the ability of leukocytes to fight infection decreases.

Future Directions in Research

More research is underway to better understand whether or not there is a standardized protein amount needed per day. Because of some difficulties in scientific techniques used to measure the increase in daily protein needs from a sedentary person to an athlete, the increase in protein needed is therefore not universal. Dr. Bob Lanier, a professor at Jesuit College Preparatory School of Dallas, writes in his discourse on minorities in developed countries that protein deficiency is more common in the developed world than has been estimated. He provides a variety of data to theorize that African American low IQ scores are a consequence of protein deficiency in the developed world. Finally, the effects of increased protein intake, both animal and plant, should be furthered tested with respect to increasing osteoporosis risk.

ESSENTIAL FATTY ACID

Essential fatty acids, or EFAs, are fatty acids that humans and other animals must ingest because the body requires them for good health but cannot synthesize them. The term "essential fatty acid" refers to fatty acids required for biological processes, and not those that only act as fuel. Only two EFAs are known for humans: alpha-linolenic acid (an omega-3 fatty acid) and linoleic acid (an omega-6 fatty acid). Other fatty acids that are only "conditionally essential" include gamma-linolenic acid (an omega-6 fatty acid), lauric acid (a saturated fatty acid), and palmitoleic acid (a monounsaturated fatty acid).

When the two EFAs were first discovered in 1923, they were designated Vitamin F. In 1930, work by Burr, Burr and Miller on rats showed that the two EFAs are better classified with the fats than with the vitamins.

Functions

The biological effects of the ω-3 and ω-6 fatty acids are mediated by their mutual interactions.

In the body, essential fatty acids serve multiple functions. In each of these, the balance between dietary ω-3 and ω-6 strongly affects function.

- They are modified to make
  - the classic eicosanoids (affecting inflammation and many other cellular functions)
  - the endocannabinoids (affecting mood, behavior and inflammation)
  - the lipoxins from ω-6 EFAs and resolvins from ω-3 (in the presence of aspirin, downregulating inflammation.)
  - the isofurans, neurofurans, isoprostanes, hepoxilins, epoxyeicosatrienoic acids (EETs) and Neuroprotectin D
- They form lipid rafts (affecting cellular signaling)
- They act on DNA (activating or inhibiting transcription factors such as NF-êB, which is linked to pro-inflammatory cytokine production).

Nomenclature and Terminology

Fatty acids are straight chain hydrocarbons possessing a carboxyl (COOH) group at one end. The carbon next to the carboxylate is known as α, the next carbon β, and so forth. Since biological fatty acids can be of different lengths, the last position is labelled as a "ω", the last letter in the Greek alphabet. Since the physiological properties of unsaturated fatty acids largely depend on the position of the first unsaturation relative to the end position and not the carboxylate, the position is signified by (ω minus n). For example, the term ω-3 signifies that the first double bond exists as the third carbon-carbon bond from the terminal CH_3 end (ω) of the carbon chain. The number of carbons and the number of double bonds is also listed. ω-3 18:4 (stearidonic acid) or 18:4 ω-3 or 18:4 n–3 indicates an 18-carbon chain with 4 double bonds, and with the first double bond in the third position from the CH_3 end. Double bonds are *cis* and separated by a single methylene (CH_2) group unless otherwise noted.

Essential Fatty Acids

Mammals lack the ability to introduce double bonds in fatty acids beyond carbon 9 and 10, hence ω-6 linoleic acid (18:2,9,12), abbreviated LA, and the ω-3 linolenic acid (18:3,9,12,15), abbreviated ALA, are essential for man in the diet. In humans, arachidonic acid (20:4,5,8,11,14) can be synthesized from LA by desaturation and chain elongation (though some carnivores like cats cannot do this, and require arachadonate in the diet). In addition, the human body can make some long-chain ω-3 PUFAs (EPA and DHA) from the ω-3 ALA.

Between 1930 and 1950, arachidonic acid and linolenic acid were termed 'essential' because each was more or less able to meet the growth requirements of rats given fat-free diets. Further research has shown that human metabolism requires both ω-3 and ω-6 fatty acids. To some extent, any ω-3 and any ω-6 can relieve the worst symptoms of fatty acid deficiency for its class. Particular fatty acids are still needed at critical life stages (e.g. lactation) and in some disease states. In nonscientific writing, common usage is that the term *essential fatty acid* comprises all the ω-3 or-6 fatty acids. Conjugated fatty acids like calendic acid are not normally considered essential. Authoritative sources include the whole families, without qualification. Traditionally speaking the LC-PUFA are not essential. See (Cunnane 2003) for a discussion of the current status of the term 'essential'. Because the LC-PUFA are sometimes required, they may be considered "conditionally essential", or not essential to healthy adults.

A 2005 study has shown evidence that the ω-6 fat gamma-linolenic acid, GLA has been shown to inhibit the breast cancer promoting gene of Her2/neu.

Essential fatty acids should not be confused with essential oils, which are "essential" in the sense of being a concentrated *essence*.

Food Sources

Almost all the polyunsaturated fat in the human diet is from EFA. Some of the food sources of ω-3 and ω-6 fatty acids are fish and shellfish, flaxseed (linseed), hemp oil, soya oil, canola (rapeseed) oil, chia seeds, pumpkin seeds, sunflower seeds, leafy vegetables, and walnuts. Essential fatty acids play a part in many metabolic processes, and there is evidence to suggest that low levels of essential fatty acids, or the wrong balance of types among the essential fatty acids, may be a factor in a number of illnesses, including osteoporosis.

Plant sources of ω-3 contain neither eicosapentaenoic acid (EPA) nor docosahexaenoic acid (DHA). The human body can (and in case of a purely vegetarian diet often must, unless certain algae or supplements derived from them are consumed) convert α-linolenic acid (ALA) to EPA and subsequently DHA. This however requires more metabolic work, which is thought to be the reason that the absorption of essential fatty acids is much greater from animal rather than plant sources.

The IUPAC Lipid HandbookPDF (370 KiB) provides a very large and detailed listing of fat contents of animal and vegetable fats, including ω-3 and-6 oils. The National Institutes of Health's EFA Education group publishes 'Essential Fats in Food Oils.'

This lists 40 common oils, more tightly focused on EFAs and sorted by n-6:3 ratio. Stuchlik and Zak, 'Vegetable Lipids as Components of Functional Food'PDF (139 KiB) list notable vegetable sources of EFAs as well as commentary and an overview of the biosynthetic pathways involved. Users can interactively search at Nutrition Data for the richest food sources of particular EFAs or other nutrients. Careful readers will note that these sources are not in excellent agreement. EFA content of vegetable sources varies with cultivation conditions. Animal sources vary widely, both with the animal's feed and that the EFA makeup varies markedly with fats from different body parts.

Human Health

Almost all the polyunsaturated fats in the human diet are EFAs. Essential fatty acids play an important role in the life and death of cardiac cells.

Essential Fatty Acid Deficiency

Essential fatty acid deficiency results in a dermatitis similar to that seen in zinc or biotin deficiency.

Treatment for Depression

Research suggests that high intakes of fish and omega-3 fatty acids are linked to decreased rates of major depression. Omega-3 fatty acids, such as docosahexaenoic acid (DHA) and eicosapentaenoic acid (EPA) are important for enzymatic pathways required to metabolize long-chain polyunsaturated fatty acids (PUFAs). Low plasma concentrations of DHA predict low concentrations of cerebrospinal fluid 5-hydroxyindoleacetic acid (5-HIAA). It is found that low concentrations of 5-HIAA in the brain is associated with depression and suicide.

There are high concentrations of DHA in synaptic membranes of the brain. This is critical for synaptic transmission and membrane fluidity. The omega-6 fatty acid to omega-3 fatty acid ratio is important to avoid imbalance of membrane fluidity. Membrane fluidity affects function of enzymes such as adenylate cyclase and ion channels such as calcium, potassium, and sodium, which in turn affects receptor numbers and functioning, as well as serotonin neurotransmitter levels. It is evident that western diets are deficient in omega-3 and excessive in omega-6, and balancing of this ratio would confer numerous health benefits.

Although further research is needed, there are studies providing evidence for the role of omega-3 fatty acids in the treatment of depression during the perinatal period. Correlations have been found between depression and low levels of omega-3 fatty acids, and treatment with omega-3 supplementation shows benefit for depression as well as other mood disorders. Research also suggests that supplementation is beneficial for healthy infant development.

Bibliography

Alka Pareek: *Environment and Nutritional Disorders*, Aavishkar Pub, Delhi, 2003.

Anderson, J. and K. Petre. *Nutrition Which Promotes Health and Wellness*. Mary Scott Lecture Series Proceedings. 1988.

Anderson, W.T.: *Evolution Isn't What it Used to Be. The Augmented Animal and the Whole Wired World*, W.H. Freeman Co., New York, 1996.

Ballentine R. : *Diet and Nutrition: A Holistic Approach*, Himalyan Press, Delhi, 1989.

Boik J. : *Cancer and Natural Medicine: A Basic Text Book of Basic Science and Clinical Research*, Oregon Medical Press, 1996.

Bower, B.: *Fossils may Clarify Mammal Evolution*, Science News, 1984.

Burkitt D.P. : *Western Diseases: Their Dietary Prevention and Reversibility*, Humana Press, 1994.

Butler, P. M., and K. A. Joysey : *Development, Function and Evolution of Animal Teeth*, Academic Pr., New York, 1978.

Campion K. : *Handbook of Herbal Health*, Sphere, 1985.

Carper J. : *Food Your Miracle Medicine*, Pocket Books, 2000.

Carroll, R. L.: *Vertebrate Paleontology and Evolution*. W. H. Freeman and Co., New York, 1988.

Conton L: *Social, Economic and Ecological Parameters of Infant Feeding in Usino*, Oceania, New Guinea, 1985.

Culton S. : *The Psychology of Stress and Nutrition*, National Book Company, 1991.

Daniel, J.C. Jr. : *Methods in Mammalian Reproduction*, Orlando, FL, Academic Press, 1978.

Daphne C. Elliott: *Biochemistry and Molecular Biology*, Oxford University Press, Delhi, 2005.

Davies S., Stewart A. : *Nutritional Medicine*, Pan Books, 1987.

Dawkins, R.: *River Out of Eden. A Darwinian View of Life*, Basic Books, Harper Collins. New York, 1995.

Eaton S. : *The Stone Age Health Programme: Diet and Exercise as Nature Intended*, Angus and Robertson, 1989.

Fan, X. : *Food Irradiation Research and Technology*, Blackwell Publishing, Ames, IA, 2006.

Ferro-Luzzi G: *Food Patterns and Nutrition in French Polynesia*, Oceania, Polynesia, 1962.

Gay, W.I. : *Methods of Animal Experimentation*, Academic Press, New York, 1965.

Gordon, G. A. : *Animals Physiology*, Harper and Row, New York, 1989.

Greene, H. W.: *Mode of Reproduction in Lizards and Snakes of the Gomez Farias Region*, Tamaulipas, Mexico. Copeia, 1970.

Greer, A. E.: *Notes on the Mode of Reproduction in Anguid Lizards*, Herpetologica, 1967.

Griffin, D.R.: *Animal Minds*, University of Chicago Press. Chicago, 1992.

Grzimek, B.: *Grzimek's Animal Life Encyclopedia*, McGraw Hill, New York, 1989.

Hill, R. W., and G. A. Gordon. : *Animal Physiology*, Harper and Row, New York, 1989.

James W.P.T. : *Healthy Nutrition: Preventing Nutrition Related Diseases in Europe*, WHO Regional Publications, 1988.

Jasra, O.P. : *Environmental Biochemistry*, Sarup & Sons, Delhi, 2002.

Lata Bhattacharya: *Animal Biochemistry*, Discovery, Delhi, 2010.

Lehninger, A. L.: *Biochemistry*, New York, Worth and Co., 1970.

Leverkus C. : *The Great British Diet*, The British Dietetic Association, 1985.

Patterson, Charles : *Eternal Treblinka: Our Treatment of Animals and the Holocaust*, New York: Lantern Books, 2002.

Prasad, S.K.: *Biochemistry of Proteins*, Discovery, Delhi, 2010.

Raghunath Narvekar: *Handbook of Biochemistry*, Adhyayan, Delhi, 2008.

Richard Myers: *Basics of Chemistry*, Atlantic, Delhi, 2007.

Robbins, John : *The Food Revolution: How Your Diet Can Help Save Your Life and Our World*. Berkeley, California: Conari Press, 2001.

Rody N: *Journal of Nutrition Education*, Oceania, Micronesia, 1978.

Roe, Daphne. *Geriatric Nutrition*. Prentice Hall Publ. 1999.

Samuel Delvin: *Science and Technology : Biochemistry*, Sarup, Delhi, 2007.

Sarath Chandra Bose: *Biochemistry : A Practical Manual*, Pharma Med Press, Delhi, 2010.

Seidel, S.M.: *New Technologies in Animal Breeding*, Orlando, FL, Academic Press, 1981.

Singh, Mahinder: *A Textbook of Biochemistry*, Dominant Pub, Delhi, 2011.

Suchman, M.: *Envisioning Life on the Nano-Frontier*, Dordrecht, Kluwer, 2001.

Sunetra Roday: *Food Science and Nutrition*, Oxford University Press, Delhi, 2007.

Sushil Chauhan: *Experiments in Environmental Chemistry*, Campus Books, Delhi, 2010.

Temple N.J., Burkitt D.P. : *Western Diseases: Their Dietary Prevention and Reversibility*, Humana Press, 1994.

Truswell A. : *ABC of Nutrition*, British Medical Journal Books, 1993.

Valli Manickam: *A Textbook of Analytical Chemistry*, PharmaMed Press, Delhi, 2006.

Verma, S.R. : *Nature: Fish Genetics and Biodiversity Conservation*, Conservators, Delhi, 1998.

Waterlow, J.C. : *Protein Energy Malnutrition*, London, UK, 1992.

Werbach M.R. : *Nutrition Influences on Mental Illness*, Third Line Press, 1999.

White, M.J.D.: *Animal Cytology and Evolution*, Cambridge, Cambridge Univ. Press, 1954.

Wurtman J.J. : *The Serotonin Solution*, Fawcett Columbine, New York, 1997.

Yablokov, A.: *Variability of Mammals: Moscow*, USSR, Nauka Publishers, 1966.

Yadav, Manju: *Mammalian Development*, Discovery Publishing House, Delhi, 2008.

Index

P

R

S

T

V

W

❑❑❑